Routing Protocols for Mobile Ad Hoc Networks

Daniel Lang

Routing Protocols for Mobile Ad Hoc Networks

Classification, Evaluation and Challenges

VDM Verlag Dr. Müller

Imprint

Bibliographic information by the German National Library: The German National Library lists this publication at the German National Bibliography; detailed bibliographic information is available on the Internet at http://dnb.d-nb.de.

Cover image: www.purestockx.com

Publisher:
VDM Verlag Dr. Müller Aktiengesellschaft & Co. KG , Dudweiler Landstr. 125 a, 66123 Saarbrücken, Germany,
Phone +49 681 9100-698, Fax +49 681 9100-988,
Email: info@vdm-verlag.de

Zugl.: Muenchen, TUM, Dissertation, 2006

Produced in USA and UK by:
Lightning Source Inc., La Vergne, Tennessee, USA
Lightning Source UK Ltd., Milton Keynes, UK
BookSurge LLC, 5341 Dorchester Road, Suite 16, North Charleston, SC 29418, USA

ISBN: 978-3-8364-6908-1

Contents

Chapter 1

Introduction

This chapter will introduce the reader to mobile ad hoc networking in general, provide background on the nature and problems of this type of networking and give an overview of the current state of research. Further it will point out the problem of evaluating routing protocol designs for mobile ad hoc networks, which is the main motivation for this book.

1.1   Commonly used terms

Since this book uses a set of apparently similar terms in different contexts, which may confuse the reader, the most important terms are briefly defined. There is a more comprehensive list of terms and their definitions in appendix A.

Application Scenario: An *application scenario* describes the characteristics of an intended application for a mobile ad hoc network including its *movement* and *traffic scenario* (cf. appendix A.2). Sample application scenarios are described in chapter 5.

Characteristic Commonly used to describe a particular, measurable *characteristic* of either a →*simulation scenario*, an →*application scenario* or a routing protocol.

Evaluation Used in the context of a →*simulation scenario* based →*experiment*, or for a benchmark comparison of simulation scenarios with the aim to establish if some of the examined items (scenarios, routing protocols) are performing in a better way than others.

Experiment The term experiment is used to describe the whole process of creating a →*simulation scenario*, actually performing the simulations and deriving results. It is also used for real world (i.e. not simulated) experiments, e.g. the CMU testbed[111].

Scenario: There are →*simulation scenarios* and →*application scenarios* (also *movement scenarios* and *traffic scenarios* which are explained in appendix A.2).

Simulation Scenario: A *simulation scenario* describes the characteristics of a simulation experiment including its →*movement* and →*traffic scenario* (cf. appendix A.2). Simulation scenarios are examined in chapter 4.

1.2   MANETs - A self-organizing networking concept

Mobile ad hoc networking is a means of communication, which does not rely on any existing infrastructure, such as dedicated routers, transceiver base stations or even cables. A definition can be found in [106].

A "mobile ad hoc network" (MANET) is an autonomous system of mobile routers (and associated hosts) connected by wireless links - the union of which form an arbitrary graph. The routers are free to move randomly and organize themselves arbitrarily; thus, the network's wireless topology may change rapidly and unpredictably. Such a network may operate in a stand-alone fashion, or may be connected to the larger Internet.

The current charter of the MANET IETF working group[107] no longer lists a specific definition, reflecting the fact that a wide range of communication forms can be considered a MANET.

Mobile ad hoc networks have been the subject of a great deal of recent and challenging research efforts[1] by many excellent scientists and Internet pioneers.

These have ranged from from very general to very special issues, covering any network layer from the physical media characteristics up to security protocols and service location.

Mainly driven by military research in the past, MANETs are about to enter the commercial platform as well, as they contribute to general scientific work. This observation is based on the growing interest in research and the existence of the first commercial techniques on the market, which work in an

---

[1]Around 1994 the first papers appeared about this modern type of ad hoc networking; although related efforts have been done much earlier, like the Packet Radio Network (PRnet) project of the US army.

ad hoc fashion. Many problems still need to be solved. Establishing and maintaining data connections for various applications between mobile nodes without any given infrastructure or even reliable cooperation, is a complex task that cannot be solved in a general way. The amount of proposed routing protocols and algorithms for MANETs reflects this nature of the problem.

1.3 Problems with Mobile Ad Hoc Networks

Multi-hop routing in such an environment is a much more complex task than routing in conventional (static) networks. This requires that all characteristics of the task will be considered. These are determined by the characteristics of the media, the behavior of nodes in terms of movement (mobility patterns) and in terms of communication (data and traffic patterns).

- Cooperation between nodes is strongly desired and may need to be encouraged.

- The used transmission medium results in comparatively low bandwidth and a high potential of channel contention.

- Due to the mobility, the links between the nodes are dynamic and can be short-living.

- The set of nodes is not fixed, since nodes may leave and join the network.

- Traffic requirements may be diverse and quickly changing.

Routing is one of the most discussed areas in mobile ad hoc networking and has generated a plethora of suggested solutions (routing protocols and algorithms). As there are many different ways to address the problems, many aspects upon which routing decisions may be based, many constraints to consider, and many different applications and performance measures to be used, over 30 different routing protocols for MANETs have been designed and presented so far. This number illustrates the variety of problems to cope with. Many are still part of research and may not be usable for a real application.

Each proposed routing method focuses on particular characteristic of the network and tries to improve things in a certain direction. As the nature of the problem is complex and the goals may even be contradictory, it is clear that a simple solution cannot be the answer.

As each proposed strategy has to be justified, simulations are commonly used to show the advantages and attributes of those suggestions. It goes without saying, that simulations that are part of a proposal of a certain algorithm, are often in favor of that algorithm.

1.4   Applications for MANETs

Mobile ad hoc networks can be applied to a large variety of use cases, where conventional networking cannot be applied, because of difficult terrain, lacking cost-effectiveness or other reasons. Examples of such situations are: A disaster area, where any possible infrastructure has been destroyed, or a military unit in the field. Other possible uses are so called *sensor networks*, intelligent sensors that transmit their data in an ad hoc manner, are deployed in an unaccessible area, Since such functionality is important and useful for the military, related organizations do fund a large fraction of ad hoc networking research.

Last but not least, there is the vision, that any mobile device that people are carrying in their everyday life (cell phone, organizer, notebook), could be MANET enabled, allowing personal communication without the need of a communication carrier.

Presently, even those example applications for mobile ad hoc networks are not closely related in terms of requirements for ad hoc routing. Consequently not all proposed routing protocols will be usable for just any purpose or for a general purpose. It is evident that future research will need to evaluate routing methods in an application scenario context. In chapter 5 I suggest a set of sample applications which are then matched against the simulation scenarios of past experiments.

1.5   Evaluation and Scenarios

Since research in ad hoc networking has resulted in a such a large amount of routing algorithms and protocols, it has become more and more difficult to decide which algorithms are superior to others under what conditions. For a successful deployment, this is an important consideration, since a wrong choice may have a severe impact on the performance, and consequently on the acceptance of the new technology. Also providing just any protocol is not feasible, due to the different requirements on hardware and lower network layers. Further, it would not make sense, since all devices in an area would need to agree on one method if they want to communicate.

Usually proposed routing protocols are justified by some kind of evaluation in the proposing paper. In most cases, there is some comparison to one or more other routing protocols using an artificial scenario, where the new routing protocol performs better according to some measure.

The used scenarios are very simple in most cases and do not model realistic conditions or certain applications. As mentioned before, the environment of an ad hoc network has a considerable impact on the required characteristics and strategies. This is valid for the real use of an ad hoc network, as well as for a simulation to evaluate a routing strategy. Since evaluations are crucial for further research and development and for a real use, special care must be taken to model the environment for evaluations so that it matches the intended use case as close as possible.

The environment consists of many aspects like physical characteristics of the medium and interfaces, type and characteristics of traffic to be transported in the ad hoc network, also movement and behavior of nodes (communication devices), and characteristics of the area, in which the nodes move.

In chapter 4, I will examine the simulation scenarios used in past and present experiments evaluating routing protocols. I classify these scenarios, compare them according to a benchmark scenario (not a concrete scenario, more benchmark constraints) and match them to the intended applications as described in chapter 5.

1.6 Status of IETF development efforts

The MANET technology has drawn so much interest of the public research and engineering community, that the MANET IETF working group was formed. This working group has submitted a large amount of Internet drafts concerning MANET routing protocols [63, 62, 84]. A drafts concerning DSR[88] has been submitted to the IESG for publication as an experimental RFC. The specifications of AODV, OLSR and TBRPF have already been accepted as experimental RFC. AODV is now specified in RFC 3561[125], OLSR in RFC 3626[38] and TBRPF in RFC 3684[117].

1.7 Outline of this Book

The order of chapters throughout the book more or less reflects the logical steps I took during my research. The description of the problems and the motivation of evaluation of routing protocols for mobile ad hoc networks in chapter 2 comes first.

Then, in the following chapter 3, I will describe the methodology that I used during my research to compare simulation based evaluations, to map application scenarios to simulation scenarios and to compare routing protocols.

I first wanted to know about how routing protocols were examined and compared in previous research. Chapter 4 examines previous simulation based evaluation experiments in the literature and provides a detailed analysis and classification of the various simulation scenarios as well as their strengths and weaknesses. The scenarios are also compared against a benchmark and compared against each other to identify similarities and to establish which classes of scenarios have been covered and which not.

After I analysed these methods, I had doubts about the applicability of these simulation scenarios to real world applications and created the set of sample applications and matched the simulation scenarios to application scenarios. I present a representative set of sample applications for MANETs along with their characteristic constraints in chapter 5. The set of applications is then mapped to the scenarios from chapter 4 in order to show the applicability of these experiments. This comparison demonstrates how well previous simulation scenarios represent certain intended applications.

Other ideas and considerations concerning models and scenarios are discussed in chapter 6.

Since I had the most important tools together, I went to do a comprehensive classification of the routing protocols themselves. The routing protocols are compared and characterized in chapter 7 with more detailed descriptions provided in chapter 8.

And chapter 9 proposes a set of guidelines for better simulation based experiments based on the observations from the previous chapters. This exposed the problems with efficient simulations and thus, chapter 10 explores the problems of using simulation as a method of evaluation itself, and presents solutions to some of the problems with simulation based evaluation by means of a framework.

Finally, a set of sample simulations is performed and analysed in chapter 11 with the aim to confirm the assumptions and conclusions from the previous chapters (e.g. if realistic scenarios yield different results than simple ones). The work is summarized and a conclusion is drawn in chapter 12.

Terms and definitions are explained in appendix A.

Chapter 2

Problems and Motivation

With the current proliferation of research directions, proposed algorithms and protocols, it is difficult for the research community and industry to agree upon a set of algorithms and protocols that can be implemented and used in the most common applications. In my opinion, one of the key problems is the lack of thorough and comprehensive evaluation. Criteria, without which it cannot be determined which algorithm will work best under what conditions and so research cannot be focused on the most promising algorithms.

Evaluation is demanding, particularly where there are so many routing algorithms and protocols, so many different performance measures and so many different application scenarios.

In an effort to improve this situation, I will examine and classify evaluations of mobile ad hoc network routing protocols in past and present research. I will propose a set of representative application scenarios for mobile ad hoc networks and provide detailed characteristics of their associated environments. Additionally, I will examine and characterize a large (but not entirely comprehensive) set of proposed routing protocols. The protocols will be compared, to identify similar approaches and also to establish what kind of protocols may be appropriate for what kind of application.

Finally, I will propose methods and describe tools developed under my supervision for the purpose of designing and implementing effective evaluation studies.

2.1 Evaluation in Past and Present Research

The vast majority of evaluation is done by simulation, the rest is performed through analytical studies and live testbeds.

A typical simulation based evaluation consists of:

- some performance measures

- a simulation scenario

- simulation software

- actual simulation runs

- processing of simulation results

- rating of results according to the performance measures

In most published simulations, the proposed routing protocol was compared to one or two others, with some independent evaluations comparing as many as four or five different protocols the most.
Each component of the simulation based evaluation process is explained below. Simulation scenarios are discussed in chapters 4 and 9.

2.1.1   Performance Metrics

Since the goal of simulation based evaluations is to determine which routing strategy *performs* best, under which conditions, the term *performance* is used very often.
However, it is not possible to define a single performance criterion and sometimes individual criteria can even be contradictory. I will describe the most commonly used performance metrics now.
Some terms are even used differently in different papers, e.g. **throughput** and **overhead** can be measured in terms of *Bytes* or *Packets*. More detailed explanations about and defitions of such terms can be found in appendix A.2.

Hop count: Also called **path length**. The number of hops, a packet has to travel to reach it's destination. In general the rule applies that fewer hops are better, as the latency is reduced and there is less potential for channel contention.

End-to-End delay: The time interval starting when the whole packet was sent until when the whole packet was successfully received at the destination.

Goodput: The amount of payload data that is transmitted in a certain time period. See also *throughput*.

Overhead: In general *overhead* is the amount of data transmitted, which is no payload data. There are many different types of overhead. Overhead is usually calculated as the difference between total data transmitted and payload data transmitted divided by the payload. Overhead can be measured differently, e.g. in terms of bytes or in terms of packets, that are transported in excess of the payload. Which measure is more appropriate depends highly on the intended application.

Route setup delay: The delay until a node is able to actually send the packet. This is caused by a missing or stale route to the packets destination. Route discovery and setup must be finished before the packet can be sent. For on-demand protocols, this is a significant factor, while proactive protocols try to minimize this delay by constantly maintaining routes.

Routing Protocol Overhead: Overhead which results from the setting up and maintaining of routes.

Retransmission Overhead: The additional data transmitted due to retransmission of lost or garbled packets.

Suboptimal Route Overhead: This type of overhead was introduced in some papers, to reflect the overhead due to suboptimal routes (i.e. routes with a longer hop count than necessary), compared to the data transmitted if an optimal route was used. A route using more hops results in more individual transmissions leading to more overhead. Other negative side effects are more channel competition and thus an increased likelihood of packet loss.

Throughput: The amount of data transmitted by the network divided by a certain time period.

Total Overhead: This type of overhead should include every effort beyond the payload transmitted over an optimal path.

Utilization: The utilization of the available network capacity.

Delivery Ratio: The amount of packets actually delivered versus those being sent. The quality of a routing strategy can be well measured against the delivery ratio.

This is just an overview of the performance metrics to be considered and used in numerous papers. More detailed guidelines can be found in [41].

The statistical data gathered by such performance metrics can and should be varied, such that minimum, average and maximum values are considered as well. Other statistical properties, like variance, may also be of interest.

2.1.2   Simulation Software

Simulation that is used for evaluation purposes is usually performed using an appropriate simulation software package. It is important that the models supported by the simulator are accurate enough for the purpose of the simulation. There are several simulation software packages available that have been used in evaluating mobile ad hoc network routing protocols, some of these are simple, others are more sophisticated.

The simulators used are:

- NS-2

- GloMoSim

- QualNet

- OPNet

- Maisie

- MARS

Section 11.3 provides more detail concerning the advantages and disadvantages of some of these simulation software packages including the one that was chosen to perform the sample simulations presented in this book (GloMoSim).

NS-2

NS-2 is the **Network Simulator 2**[116] that emerged from the VINT[161] project at ISI[77]. It can be used to simulate any kind of Internet communication, providing implementations for IP, TCP (different flavors), UDP, a variety of routing protocols, several QoS mechanisms, and more. A project at the CMU has provided wireless extensions, which now include mobile nodes and wireless communication with adequate models on layer one and two (radio propagation models, IEEE 802.11 link layer, etc). Some ad hoc routing protocols such as AODV, DSR and TORA over IMEP have also been implemented in NS-2.

The implemented models are generally very detailed, which leads to high complexity in the software itself and in its calculations. NS-2 is written and developed in C++ and TCL. It has an embedded TCL interpreter, such that TCL scripts can be used to configure and control the simulator. The mix of C++ and TCL also increases the complexity of the software.

This is one of the main drawbacks of NS-2. Despite being open sourcem the design of NS-2 is somewhat complex, making additions or improvements difficult to implement. Given that no clear guidelines are provided, contributions have tended to add code in a haphazard manner to best suit their own purposes at that time, which has resulted in a very complex C++/TCL jumble. In recognition of these drawbacks, there have been efforts to improve the structure and design of NS-2.

NS-2 does not provide any statistics, that could be mapped to performance measures (cf. Section 2.1.1). Instead every event produced by the simulation is written to a trace-file. The trace-file can then be processed to extract the desired information. This may appear as a reasonable approach, however, even with low scale simulation scenarios the trace-files become very large. In addition, the writing of the trace-file to disk actually slows down the simulation in some cases. As an example, a wireless simulation scenario with ca. 40 mobile nodes and a simulation time of 1500 seconds creates a 2GB trace-file and takes around 45 minutes on a Sun Ultra 2.

Still NS-2 is one of the most widely used simulators for mobile ad hoc networks and there is also an add-on called NAM (network animator), which provides a way to visualize simulated communications (and NAM was designed to operate on a particular kind of trace-file, thus for NAM trace-files are always required).

GloMoSim

GloMoSim[158, 159] is a simulation environment based on a simulation package called *PARSEC* [122]. GloMoSim has a layered model of network communication according to the ISO/OSI network model. An additional module deals with mobility. GloMoSim is written in C with PARSEC extensions. The GloMoSim code is freely available, but PARSEC itself is not, however, PARSEC is distributed along with GloMoSim (in binary form).

GloMoSim has a straightforward configuration file, which provides a large set of variables to specify the simulation properties. This simpler approach is not as powerful as the TCL language from NS-2, but it nevertheless it proved to be sufficient for the simulation and reduces configuration complexity and confusion. Statistic values are collected and printed after the simulation is done, which is another advantage of GloMoSim.

Despite some considerable disadvantages, GloMoSim was used to perform the sample simulations of this work. The choice of GlomoSim and details of the results are explained in chapter 11.

QualNET

QualNet[141] is the commercial version of GloMoSim, which offers many advantages such as parallel execution of simulations on SMP hardware and improved support for a larger variety of MANET routing protocols. The proXXXX licensing costs of QualNet excluded consideration of more than a very limited version during the drafting of this project. QualNet has rarely been used in published simulation based evaluation of MANETs.

OPNET

OpNET Modeller[118] is another commercial network simulation software package. Like QualNet its use in past evaluations is limited, thus OPNET has not been considered for the work of this book.

MaRS

The Maryland Routing Simulator[160] was used to evaluate only a small fraction of examined routing protocols (LAR, cf. section 8.20), thus it was not part of further studies or considerations of this book.

Maisie

Maisie is a parallel computing simulation language developed by UCLA[109]. Maisie is no longer supported and has been succeeded by PARSEC[122].

Chapter 3

Methodologies and Motivation for Comparison

In chapters 4, 5 and 7 I will examine simulation scenarios, application scenarios and routing protocols for mobile ad hoc networks. In order to display relations between simulations, applications and routing protocols, comparison criteria must be established. To facilitate comparison against a common frame of reference, benchmark characteristics for the simulation scenarios are needed. Thus it is necessary to define comparison functions and further to represent the characteristics of simulation and application scenarios and the characteristics of routing protocols in a way that is suitable for such a comparison and evaluation.

3.1 Types of Comparisons and Motivation

There are several issues that can be addressed by comparison.

In chapter 4, I present previous simulation scenarios and I discuss the characteristics of these simulations with respect to applicability in terms of realistic node behaviour. To gain an overview of the coverage of the previous simulations, I am using a comparison function to establish a *Similarity-Relation*. This relation can show which pairs of simulation scenarios have more related or even shared characteristics than others. Thus a first impression is gained about the diversity of such scenarios.

A comparison against a set of benchmark characteristics (in the sense of a reference scenario) of these previous simulations is presented in the chapter 4. The reference values are chosen to represent a most "realistic" simulation environment. This base comparison is used to establish an initial but very rough ranking of which simulations are more useful than others.

This does not consider that different applications of MANETs will have different environmental constraints and therefore require a different simulation scenario, which is addressed in the following chapter 5. In this chapter, simple comparisons are avoided, instead the simulation scenarios are matched against a set of sample application scenarios. The characteristics of the applications are chosen to best describe an application environment. Since the previous simulations were not oriented towards such real applications, there is no one-to-one correspondance of simulation and application scenario characteristics. In particular the application scenarios proposed are described with increasingly more detailed characteristics. In order to still be able to match simulation scenarios to these applications, the comparison function used takes this into account. Whilst not all possible characteristics can be compared, it is still possible to compare those that are available and have a corresponding value. This can also be achieved by deriving some of the values (e.g. node density can be derived from the size of the area and the number of nodes). With this matching of simulation scenarios to sample applications, it can be established which simulation scenarios are most useful because they best represent an application, which types of applications are covered by existing simulation scenarios, and finally which may not be covered at all (and are therefore subject to future evaluations).

Comparison is also used in chapter 7, but for a different reason. In this chapter, the routing protocols are described and classified. Since there are so many routing protocols proposed with many different characteristics, it is difficult to decide, which protocols should be chosen for an application or even for the simulation of a certain application. The amount of protocols requires careful choice before any evaluation is done. To help with such decisions, I have established an extended classification and I also compare the routing protocols with each other, again resulting in a *Similarity-Relation*. This one shows which routing protocols are of a very similar design and thus possibly perform in a similar manner in the same environment. The relation of routing protocols can give hints for future development. Two very similar protocols could be developed together in the future or a single protocol could be designed using the advantages of both.

3.2 Relevance of Comparisons

Comparison of Simulation Scenarios with each other As described above, the comparison of simulation scenarios establishes a similarity relation, such that for each pair of simulation scenarios a degree of similarity is determined. The meaning of this similarity is, that two "similar" simulation scenarios share many common characteristics. First the set of characteristics for simulation scenarios is chosen. Only such characteristics are used, that can be identified for most simulation scenarios (it is still possible that the value for a particular characteristic is unknown). If all characteristics are equal a similarity value of 1.0 is achieved. Vice versa, the lowest possible similarity is 0. This relation, however, is not transitive. It is possible that two scenarios are very similar to a third scenario, but not at all similar to each other (due to different characteristics being similar).

Comparison of Simulation Scenarios against Benchmark Values The evaluation is done by using the comparison function aforementioned and some basic benchmark values (or ranges). The function is modified in order to allow a comparison against intervals instead of exact values. If all characteristics are within the specified benchmark intervals, a maximum of 1.0 can be achieved.

Comparison of Application and Simulation Scenarios In order to match simulations to application scenarios, the comparison function needs to take into account that some of the corresponding characteristics are not represented in the same way and should be adapted first. Further, other characteristics cannot be used for the matching at all, since they do not have a counterpart.

Comparison of Routing Protocols Just like the comparison of simulation scenarios, the routing protocols are placed in relation to each other, resulting in a degree of similarity. The meaning is the same, such that "similar" routing protocols share many common characteristics. The more characteristics are the same (or in the same range), the higher the similarity between these routing protocols. If a protocol is compared with itself, of course 1.0 similarity is the result. Again this is no transitive relation.

3.3 Representation of Characteristics

In order to be able to compare simulation scenarios with each other or with application scenarios, I have selected a set of *characteristics*. These characteristics represent the key aspects of each simulation scenario and application scenario. For the comparison of routing protocols, representing characteristics are used. The characteristics are represented in a way that allows comparison using a computer program.

I have chosen a simple and straightforward approach, which is inspired by the data representation in genetic algorithms.

First I have established the set of key characteristics of each simulation scenario and my sample application scenarios, which are described in sections 4.1 and 5.2.

In section 7.4 I have also established the characteristics of routing protocols The representation of these characteristics can be described as a sequence of elements. Each element of the sequence represents one characteristic of the simulation scenario, application scenario or routing protocol.

3.3.1 Encoding Method for Characteristics

Any approach that relates certain characteristics to others in a scientific manner (qualitatively or quantitatively) will require some sort of formal representation of these characteristics.

This formalization of the characteristics can be a very difficult task. The most promising idea was to use a sequence code for the characteristics (as used in genetic algorithms) and to define a comparison function for each characteristic.

In general I will use the following outline for the sequence code:

- Each routing protocol, scenario or application is described by a sequence of a fixed number of characteristics, called *elements* of the sequence, E_i. The position i of an element in the sequence is called it's *type*.

- Each *element* consist usually of a string of letters or a number.

- The *elements* are separated by a special character, I use the colon ":".

- Functions to compare elements of the same *type* are defined.

- The special character "?" is allowed in elements, representing an unknown value. The unknown value will be handled separately in the comparison functions.

The sequence can be written as follows:

$$S = < E_1 : E_2 : ... : E_n >$$ (3.1)

with

$$E_x = < C_{x_1}, C_{x_2}, ... C_{x_k} >; C_{x_y} \in A_x$$ (3.2)

$(C_{x_1}, ... C_{x_k}$ are letters from alphabet $A_x)$

or

$$E_x \in \mathbb{R} \tag{3.3}$$

for each type x.

Thus S is a sequence of strings E_x which in turn consists of characters C_{x_y} each from alphabet A_x or real numbers. This form still allows a flexible definition of each element for each type to adequately represent the corresponding characteristic.

3.3.2 General Description of the Comparison Functions

For the comparison functions, each character of each element of one scenario (or routing protocol) is compared with the characters of the corresponding element of the other scenario (or routing protocol, application scenario, etc). The comparison function generally used where a string of letters is representing the characteristic is as follows (with $\mathbb{E}$ being the set of all possible strings E_x).

$$f_x : \mathbb{E} \times \mathbb{E} \mapsto [0:1] \tag{3.4}$$

$$f_x(E_x, E'_x) = \begin{cases} 1 & \text{if} & E_x = E'_x \\ 0.5 & \text{if} & E_x = \text{``?''} \wedge E'_x \neq \text{``?''} \vee E'_x = \text{``?''} \wedge E_x \neq \text{``?''} \\ 0.25 & \text{if} & E_x = \text{``?''} \wedge E'_x = \text{``?''} \\ 0 & \text{else} \end{cases}$$

$$\tag{3.5}$$

As described in 3.5, an exact match yields 1.0 for that part of an element. No match results in 0.

Since it is possible that the value of some element is not known for a certain scenario, the following rule applies to such comparisons:

- If only one of two values to be compared is unknown, the individual comparison value will be set to 0.5.

- If both values to be compared are unknown, the individual comparison value will be set to 0.25 (i.e. 0.5^2).

I have chosen this approach to deal with unknown values because this prevents unknown values disturbing the result as it would be the case if no matching were be assumed, or if this individual comparison were ignored.

Of course, 3.5 describes only special cases of the comparison functions. In the general case, the comparison functions can be much more complex, e.g. to yield values > 0 for comparisons where characteristics are not the same (i.e. no equality), but still similar.

Some of the more complex comparison functions use multiple sub-functions, e.g., to compare occurrences of letters in the sets of letters representing a characteristic. These sub-functions may be weighted. Such functions yield discrete results.

In most cases where a number is used to represent a characteristic, a continuous (or partially continuous) function can be used.

The most common function to compare numbers is as follows:

$$f_x : (\mathbb{R} \cup \{?\}) \times (\mathbb{R} \cup \{?\}) \mapsto [0:1] \tag{3.6}$$

$$f_x(E_x, E'_x) = \begin{cases} 1 & \text{if } E_x = E'_x \\ 0.5 & \text{if } E_x = \text{``?''} \wedge E'_x \neq \text{``?''} \vee E'_x = \text{``?''} \wedge E_x \neq \text{``?''} \\ 0.25 & \text{if } E_x = \text{``?''} \wedge E'_x = \text{``?''} \\ \frac{E_x}{E'_x} & \text{if } \frac{E_x}{E'_x} < \frac{E'_x}{E_x} \\ \frac{E'_x}{E_x} & \text{if } \frac{E_x}{E'_x} > \frac{E'_x}{E_x} \end{cases} \tag{3.7}$$

Usually the numbers are compared in terms of a fraction. The smaller fraction of the two values is used as the result to ensure that the result is between 0 and 1. The case in which E_x or E'_x is equal zero (und thus potentially requiring division by zero) does not occur.

3.3.3 Comparison Result

In the previous section, I have described the functions to compare individual characteristics (the elements of the sequence). The overall comparison result is then computed as a weighted average ($\mathbb{S}$ is the set of all possible sequences S):

$$f : \mathbb{S} \times \mathbb{S}' \mapsto [0:1] \tag{3.8}$$

$$f(S, S') = \frac{\sum_{i=1}^{|S|} w_i f_i(E_i, E'_i)}{\sum_{i=1}^{|S|} w_i} \tag{3.9}$$

In some comparisons, the individual comparison function for a characteristic has an associated *weight* w_i. The overall result of a comparison is the sum of the comparison results for each characteristic, each multiplied by its associated weight and divided by the sum of all weights.

3.4 Comparisons and Evaluations not done

As a simulation based experiment is just a sample, more generally valid results might be achieved by analytically evaluating each routing protocol in each application scenario.

Unfortunately it is very difficult, if not impossible to define an analytic evaluation function for each routing protocol in the environment of an application scenario. The only practical way to undertake such an evaluation is by simulation, but to achieve confident results in the *Routing Protocol × Application* space, a very large amount of simulation experiments would need to be done. This work suggests ways to reduce this large number and thus to facilitate future more comprehensive investigations. A simulation based on a few sample scenarios and routing protocols is presented in chapter 11

3.5 Implementation of Comparisons

The comparisons and evaluations have been implemented using the *Perl* scripting language [162]. A small perl program has been developed for each comparison performed. The program reads the data in a structured field representation, which is attached to the perl code, parses it and calls corresponding comparison or evaluation functions for each field value.

Chapter 4

Scenarios Used in Previous Evaluations

This chapter will first introduce the common characteristics of the simulation scenarios used in various published evaluations. The scenarios are then classified and compared according to these characteristics so that the common types of simulation based evaluation are identified.

4.1 Characteristics and Quality of a Scenario

In this section I will present the key characteristics of a scenario in detail, and their importance for the quality of that scenario.

Observed Area As defined in section A, this is one fundamental characteristic. It is a mandatory characteristic, and no scenario based simulation[1] can be performed without it. Its two main parameters are **shape** and **size**. The **shape** used in any scenario for the evaluation of ad hoc network routing protocols is a *rectangle*. There are good reasons for that: A rectangle is easily specified, most simulation software only supports rectangles. Other shapes do not offer any obvious advantage. The **size** does vary much more and is indeed a parameter that affects the real world situation to be modeled very much. Sizes vary from a small room (3m × 6m) to an area that could cover several towns (10000m × 10000m).

Good quality scenarios can be of any size, but the size should reflect the intended use case as close as possible. Usually this is not a problem.

---

[1]Of course, complexity analysis or other analytical work does not need a scenario at all and could be performed without a fixed observed area.

Types of nodes As described in section A, nodes can be of different types, which will behave differently (pedestrians will move differently from cars). Each node type can have certain characteristics itself. The level of detail for the model is reflected by the use of such fine grained characteristics. The capabilities of nodes described here do not imply an overall *mobility metric* for the scenario, since these are just possible values of a single node type.

The node type determines the following parameters, which may or may not be present in the various models:

- Likelihood of moving at a certain time
- Capabilities in terms of
  - *acceleration*
  - *deceleration*
  - *maximum speed*
  - *change of direction*
- Interaction with certain subareas (e.g. *cars* can only move on *streets*)
- Moving strategy (as explained in Section 6.3)
- Time intervals of operation (the node may be an active part in the network only during certain time intervals).

Number of Nodes This is a very basic parameter of a scenario. Both the overall number of nodes, as well as the number of nodes of each different type are important. In my opinion the overall number of nodes will have a larger impact on the simulation results, as it determines also the *node density*.

Radio Model and Radio Range The radio model should reflect the kind of radio hardware used for communication. Often this choice determines also the link layer. A large variety of hardware is available, but not much of it can be used for mobile ad hoc networking. Parameters that depend on the radio model are (among others) **channel bandwidth** and the **radio range**. In combination with *node density*, the *radio range* will have a considerable effect on the results of the simulation because it influences connectivity and channel competition, two effects of opposing benefit.

Radio Propagation and Obstacles (for signal propagation) In a real world scenario, the observed area will consist of a flat and free space only in very rare cases. More often, there will be obstacles present. There are different ways to model this. One way is to use a general radio propagation model, which statistically restricts the propagation and therefore the range of the radio signals. Another method would be, to explicitly allow the placement of obstacles in the area, that specifically reduce the range of the radio signals. The placement of *subareas* (see appendix A), with a certain characteristic that affects the propagation of a signal through this area can be used to represent this. However most simulation software does not support explicit placement of obstacles or special areas.

Restricted Areas and Obstacles (for movement) Obstacles may obstruct not only radio signals but also movement of nodes. This can be reflected in the scenario by the placement of *subareas* which have certain restrictions for node movement. E.g. nodes cannot move through a building, or nodes of type "car" can only move on subareas of type "street". So the scenario could allow the definition of types of subareas with certain characteristics, that will affect node movement. In my opinion this can be a very important element for some scenarios to provide close modelling of a real world situation.

Border Behavior This is an important, but often neglected aspect of the scenario. [15] and [29] have shown, that the border behavior has an important impact on the user distribution over the area, which affects local density and therefore the simulation results. The manner in which nodes behave on approaching a border will certainly affect the realism and the applicability of a scenario. See also section 4.5.5.

Introduction and Removal of Nodes This characteristic is related to the *border behavior*. In a real world scenario it will happen that nodes enter the observed area, while others leave it. Further, nodes within the area can be switched off and thus cease to be part of the network and at a later time they will be switched on again and resume participation in the ad hoc network. I consider the possibility to reflect such behavior in certain scenarios as an important contribution towards realism and applicability.

Group Mobility The possibility to form groups (as defined in appendix A) of nodes and the flexibility of group criteria, and the quality[2] of the

---

[2]again in terms of realism and applicability

group mobility model also contribute to my quality measure of scenarios.

Observation Time This is the duration of the scenario, which usually corresponds to the simulation time (although, it would be possible to simulate several steps of a scenario separately). I consider this not a real part of the scenario, but a simulation parameter, thus it plays only a minor role in this work. It is mentioned for the sake of completeness and because it affects the runtime of a simulation and such the confidence of the result. Very short observation times may be subject to initialization side-effects. While very long observation times in existing simulations indicate a less detailed model of other simulation aspects (e.g. no modeling of the physical properties of the wireless interface).

Mobility and Mobility Metrics It is important to derive more parameters from the given characteristics (like *node density*), that can be used as a measure of mobility or a *mobility metric* (cf. appendix A). For the comparison of routing protocols in terms of *performance* (cf. appendix A), it is essential to determine a *degree of mobility*.

The impact of *mobility* (in terms of average speed, pausing periods, direction changes, etc.) is expected to be significant for the performance of certain routing protocols. It is expected that some algorithms perform much better under *"high mobility"* than others, while with *"low mobility"* there may be no difference. It is expected that a higher degree of mobility will make it more difficult for the routing protocol to perform well.

The characteristics described above, will be used to exam existing simulation based evaluations, to identify the scenarios used, and also to categorize and rate them in terms of quality.

4.2   Currently Used Scenarios

This section will summarize the scenarios used in previous papers, categorize them and there will be also some statements about their quality (in terms of applicability).

4.2.1   Common Observations

Area: All applied scenarios have used a rectangular area. This is a sensible choice, as other geometric shapes don't offer any particular advantages.

Role of Nodes: All nodes are assumed to be devices, carried by persons or in vehicles controlled by persons. They only move on ground-level in two dimensions.

4.2.2 Simple Scenarios

I define scenarios with the following characteristics as *Simple Scenarios*:

- The observed area is a flat empty space. There are no subareas, obstacles or other movement restrictions. Nodes can move arbitrarily within the area.

- Nodes cannot leave the area, new nodes cannot be introduced, and nodes are always active.

- Nodes move according to a *simple* strategy (cf. appendix A), like *Random Waypoint* or *Random Direction*.

- Nodes can only move at a constant speed.

- Nodes don't change direction during a single move. All direction changes are *sharp*, there is no smooth turning or curves.

Scenarios like these have been used in many simulations, with some minor differences. I will now describe what variants have been used.

Basic Random Waypoint Scenarios

This type is used frequently. The *Random Waypoint* movement model implies that each node chooses a random destination within the given area, moves to that destination at constant speed on a direct path and then waits for a fixed *pause time* (cf. appendix A), before choosing the next destination. This scenario has been used in [25], [45], [73], [170] and others. In these works, further characteristics are:

Area size:	1500 × 300m and 2200 × 600m (in [45])
Radio Range:	250 m
Number of nodes:	10, 20, 30, 40, 50 and 100 (once)
Speed of nodes:	[0..1] m/s and [0..20] m/s
Pause times:	30 - 900 seconds, globally fixed for all nodes
Simulation time:	500 and 900 seconds

The authors of [86] also use a related scenario, and their work is the paper that introduced the notion of *"Random Waypoint"* mobility model. It is one of the earliest papers on the subject and the scenario itself differs a lot from those in the other papers[3]; I mention it here, but it does not really belong to this (or any other) class of scenarios.

A similar situation exists with [60], where a scenario is used with an area of 5000 × 7000m, pause times from 30 to 90 seconds and only 20 nodes.

Mobility Metric   Although not explicitly mentioned, these scenarios use either the *pause time* or the *mean speed* as a mobility metric.

These parameters are varied in the simulations to reflect different degrees of mobility. Although such a simple mobility metric cannot reflect all aspects of mobility, it appears sufficient for simple mobility models. More complex mobility metrics will be discussed in sections 4.2.3 and 4.5.2. [129] suggest a better mobility metric for the *"Random Waypoint"* mobility model, derived analytically.

Border Behavior   The movement strategy determines that there is no border behavior required. The nodes always choose a destination within the area, and the area is convex so a border is never crossed (although it may be reached).

Node behavior   Nodes cannot accelerate or decelerate. Their direction is determined by the current destination point, and the likelihood of moving is determined by the pause time (i.e. always moving, except during pause phase).

Area Size and Shape   The odd area size of 1500 × 300m (which is used widely) is argued to stress the routing protocol more than a 1000 × 1000m scenario. It allows a high node density together with long paths, without the need for many more nodes (which would lead to problems with the simulation due to the extended runtime, cf. section 4.4.1).

Modified Random Waypoint Scenarios

A modified version appears in [85] where the model was extended such that the *pause time* is not globally fixed, but can be chosen at each individual

---

[3]The ranges are much more limited, i.e. the area is a 9 × 9m room, radio range of 3m, simulation for 4000 seconds with 6 to 24nodes.

movement. The area used was 1000×1000m but the simulation ran only for 250 seconds. In this paper more sophisticated scenarios (like "disaster area", cf. section 4.2.3) have been used, as well.

In [21] there is also a modified version called *Restricted Random Waypoint*. As the modification introduces some special regions, the model is categorized as a more advanced model and described in section 4.2.3.

Random Direction Scenarios

This scenario was described in [128]. Nodes move in a certain direction within $[0..2\pi]$ with a speed of $[0..10]$ m/s until they hit the border. Then they wait a certain time, before choosing a new direction from $[0..\pi]$ relative to the "wall" (nodes are reflected from the borders). Thus, the border behavior plays an important part of the movement model itself, and is therefore defined in precise way. In this case, a contact with the border is the only reason for a node to stop. The characteristics of the random direction scenario are:

Area sizes:	1000×1000m, 1500×1500m, 2400×2400m and 3450×3450m
Radio Range:	250m
Number of nodes:	50 and 500
Speed of nodes:	$[0..10]$ m/s
Direction:	$[0..\pi]$ relative to the "wall"
Pause times:	on each border hit, but duration not specified
Simulation time:	300 seconds

The evaluation in [70] uses a similar model with nodes being reflected from the border on contact, but not pausing. The size used in [70] is 1000×1000m with 200 nodes and the radio range is 105m.

Both scenarios seem rather artificial and appear to provide the least realistic movement patterns for nodes like pedestrians, cars, bicycles, etc.

Other Simple Scenarios

In [44] a different approach is chosen. Each movement is specified by a triple of *direction*, *speed* and *distance*, which have been chosen at each step as follows:

Area size:	1000×1000m
Radio Range:	350m
Number of nodes:	30 and 60
Direction:	chosen from $[-\pi/8.. + \pi/8]$ relative to the previous direction.
Speed:	chosen from $[0.4..0.6]$m/s and $[3.5..4.5]$m/s
Distance:	exponentially distributed over a mean of 5m.
Simulation time:	10000 seconds

4.2.3   Advanced Scenarios

The following scenarios are more advanced. Some introduce obstacles (hindering both node-movement and radio propagation). There are different types of nodes with different properties possible. Certain regions within the area (subareas) are utilized and can impose certain restrictions to nodes in that subarea.

Johansson Scenarios

[85] describes three scenarios, which are very different from the simple ones, and which appear far more realistic in terms of node behavior than the simple scenarios.

They allow the use of obstacles that absorb any communication, such that no link can go through an obstacle. Alas, the movement strategy is not described in the paper.

Mobility Metric   This paper provides a much more thorough and detailed mobility metric which is consequently much more complex. The approach is general enough to be used as a basis for other scenarios. The following sketches the idea of the mobility metric:

The value of $|v(x,y,t)|$ (with v being defined as the relative velocity of nodes x and y at time t) is averaged over time and then averaged over all node pairs. For more details I refer to [85].

Conference Room   A conference room is modeled with a *speaker node*, several listeners and a few people moving around. This is a rather static scenario. Only 10% of the nodes move, with a maximum speed of 1m/s. Most nodes are assigned to specific locations, but are still able to move. It is not specified how they move. Nodes can be blocked by obstacles. There are different types of nodes: a speaker, several curious bypassers and the remaining are attending listeners. Other known parameters are:

Area size: 150×90m
Radio Range: 25m
Number of nodes: 50
Speed: < 1m/s
Simulation time: 900 seconds

Event Coverage & Disaster Area   The *Event Coverage* scenario should model a large event, like a trade fair, with several groups and individuals moving on a large area. As in the *Conference* scenario, the nodes move with 1m/s but at least 50% of the nodes are moving. There are obstacles as well, and there is some chance that up to 10 nodes may form a group. The cause and implications of such a group forming are not stated clearly, but it is likely that they move together. The movement strategy is not described at all.

The *Disaster Area* (which should resemble the site of a large accident) scenario differs only in the manner of node movement. There are three distinct areas, which nodes cannot leave and which are too far away for a direct communication. Nodes move randomly within each area. Two dedicated nodes (which should model helicopters) move between these areas with a much higher speed of 20m/s.
Parameters for both scenarios are:

Area size: 1500×900m
Radio Range: 250m
Number of nodes: 50
Speed: < 1m/s and 20m/s for 2 nodes in Disaster Area
Simulation time: 900 seconds

Restricted Random Waypoint

The *Restricted Random Waypoint* scenario used in [21] introduces *town* and *highway* regions. Within a *town* region, the usual *Random Waypoint* model (cf. section 4.2.2) is used. After a certain amount of moves, a node chooses a destination in another town. Additionally, there are *commuter* nodes, that move between the towns with a higher speed and a pause time at each town for 1 second. Areas of 3500×2500m and 4500×3500m have been simulated with three *towns*. Each town is a square of 600m side length. The following parameters have been used:

Area size:	3500×2500m and 4500×3500m
Town size:	600×600m
Radio Range:	250m
Number of nodes:	400 (100 regular, 300 commuters) and
	600 (with 500 commuters)
Speed:	< 10m/s (regular nodes) and [10..20]m/s (commuters)
Pause time:	[0..200] seconds in steps of 50 (regular nodes) and
	1 second for commuters
Steps in town:	20
Simulation time:	not specified

It is not clear, why this scenario is called *restricted*. It is possible, that the movement of the nodes can be regarded as more restricted than in the usual *Random Waypoint* model, since most nodes cannot leave a town area until they have performed a certain amount of moves, but are then forced to move to another town. Their freedom of choice is more limited in that sense.

4.2.4   Real Installations: CMU Testbed and AODV Testbed

[111] describes a testbed with a real installation of DSR [86]. The scenario consisted of 5 cars with laptops equipped with standard WaveLAN cards, as well as two fixed nodes, 750 m apart. The cars move constantly in a loop around the fixed nodes, but there is real traffic on the roads.

On March 25th and 26th 2002, a successful test of real AODV implementation using both IPv4 and IPv6 was done at the UCSB. A report about that event is available [11].

4.3   Other Models and Tools

4.3.1   Modeling Turning and Acceleration: Smooth is Better than Sharp

C. Bettstetter proposed a *smooth* mobility model in [15]. This is not a scenario description, as proposed in most other papers, but a fine grained movement model, that focuses on the kinetical characteristics of a move of each single node. It introduces acceleration and deceleration of nodes, as well as speed-correlated direction changes. To complete the model with some movement strategy, a Poisson process is assumed. It generates *speed change* and *direction change* events during the simulation time. The events are generated according to an exponential distribution, using $\lambda = p_{v^*}/\Delta t$, with p_{v^*} being the probability of a change event at each time step Δt.

So, unlike the other scenarios, the model does not assume individual discrete movements, but is driven by these speed change and direction change events.

This shows that this model was not designed with the prerequisite to work with simulation software packages which just accept constant speed movement descriptions. The common simulation tools NS-2 [1] and GloMoSim [158] have these limitations. It is still possible to derive such movements from the model, by discretizing an accelerated movement into small steps of increasing (or decreasing) constant speed. The same is possible for the turns. The accuracy then depends on the time resolution, however a high resolution will result in an increasing amount of discrete constant-speed-movements, for a single move, on each speed change or direction change event.

I have implemented the model, in order to see how well it matches collected GPS data. Due to the poor quality of the GPS data, it was not possible to draw clear conclusions, but statistical data from the implementation is presented in appendix C.

4.3.2 Scenario Generators and CADHOC

There is a small set of scenario generators available, but most of them are only capable of generating scenarios already described above, i.e. *simple scenarios*, with some minor enhancements like group mobility (e.g. scengen[130]).

The only notable exception is CADHOC[142], which is a Java based scenario generator that is capable of creating more "realistic" scenarios than other tools. The main advantage in terms of realism is that it is possible to define regions where the nodes can and cannot move. So one could create a building with rooms and halls for pedestrians, a street pattern for cars, etc. The initial location of each node can be specified, as well as movement patterns from a restricted set of strategies including a *Brownian movement* and a *pursuit model*. CADHOC is also capable of generating data traffic between the nodes.

Unfortunately, this tool is very awkward to be used efficiently, as it is GUI driven and requires a lot of resources to run. After the specification, it takes a very long time to actually create the scenario. Although the concept is promising, in practical terms it becomes unusable if you want to create many different patterns from a single scenario specification, or if you want

to specify many different scenarios and create even more unique scenarios[4]
from each specification.

4.4   Discussion: Why Have These Scenarios Been Used

Although advanced scenarios exist they have been used rarely. One would
expect much more different or more sophisticated scenarios to be utilised.
So, why are these simple models so often chosen, instead? Two main reasons
suggest themselves:

1. **Comparability:** The random waypoint scenario with 1500×300m area
   was used in very early evaluations like [25]. Subsequent developments
   and evaluations aimed to be comparable to the earlier results, such
   that a statement about the performance of the developed algorithm (or
   routing protocol) could be made. So even independent studies like [85]
   [5], that compared a whole set of routing protocols, used these scenarios.

2. **Simulation Constraints:** This may be the reason, why such a scenario was chosen in the first place. As mentioned previouslym there
   are two simulation software packages that are very commonly used in
   evaluating ad hoc networks. These are: NS-2[1] from the VINT project
   at Berkeley and GloMoSim[159] a PARSEC based simulation package
   developed at UCLA.

   Sophisticated simulation software like NS-2 and GloMoSim (both model
   all network layers in great detail) results in complex calculations. The
   computing time and memory requirements do not scale with increasing
   node numbers. This makes it difficult to simulate more sophisticated
   scenarios. In particular, NS-2 consumes a huge amount of resources for
   more than 50 nodes, and produces a huge amount of data. Simulations
   with more nodes and for a longer simulation time are nearly impossible with NS-2, even on very powerful machines. GloMoSim seems to
   perform better, but still consumes a lot of memory.

Due to the fact, that the *random waypoint* model was used in many studies,
analytical research about it was performed in [16, 129] and [17]. The result

---

[4]In this context, a unique scenario is the set of exact movement and traffic instructions
for each node at any time step.

[5]independent in the sense, that the author of the study is not also the author of a
routing protocol

concerning node distribution and increasing density in the center of the area is a particularly important fact to be aware of during evaluations using this model.

4.4.1 Why the 1500×300m area?

Section 4.2.2 already covers some possible reasons. The dimensions are chosen relative to the transmitter range, which is commonly around 250m, so that in one direction multi-hop links need to be established. As mentioned before, a higher node density in combination with the need for multi-hop paths can be achieved with a lower number of nodes. A further reason for the narrowness of the area is that this forces movements primarily in the "extended" direction, thus causing link breaks and stressing the protocol.

4.4.2 Why random waypoint/random direction ?

The *random waypoint* model maps very well to the input data NS-2 and GloMoSim require. So it is very easy to use the data of such a model with these two simulators. The model itself is also simple and therefore easy to implement. *Random direction* is equally simple and easy to map to the simulation software.

4.5 Criticism of Proposed Scenarios

4.5.1 Node Behavior

The scenarios with the most questionable behavior are clearly the *simple* scenarios. There is reasonable doubt that devices attached to people, or people operated vehicles, would move in the manner suggested by the *Random Waypoint* or *Random Direction* model. It has been established in [29], that the *Random Waypoint* model is vulnerable to some initialization problems, which leads to a very unstable neighbor set in the first $600 - 1000$ seconds of a simulation. In addition, a clear area with no obstacles or restrictions will rarely occur in a real deployment. Further the observed time intervals are rather short, although this may be acceptable for such simple scenarios, since there would not be much change over time anyway.

Among the more complex scenarios, the Johansson scenarios[85] are a big step in the right direction. The scenarios have been modeled after certain real-life situations, there are obstacles, certain restricted movements and group mobility. More investigations regarding this work would have been very interesting. Johansson et al. did announce in their paper that more work was in progress, but it seems that this was never published. The simulation itself is questionable, as it seems that for each scenario only a single simulation was performed. From a statistical viewpoint, this is certainly not adequate. This major drawback was already pointed out in [36], but since this problem is not related to the scenarios themselves, I consider them some of the more appropriate scenarios.

Apart from Johansson's evaluation and CADHOC[142], there are no restricted regions, that could induce some kind of "channeling" of the nodes or force some other kind of correlated behavior. Different kinds of nodes and group mobility are only used in rare cases, although [130] would support both. Accelerated movement is not used at all.

The recently published paper[81] comes to similar conclusions concerning the applicability of the Random Waypoint or Random Direction mobility model. In this paper an *Obstacle Mobility Model*[6] is proposed, which provides a great enhancement to the previous simple models. Since this paper was published only recently, there are no further studies using this mobility model, yet.

4.5.2   Mobility Metric

The need to define a mobility metric parameter (as described in section 4.1) is not commonly understood. The parameters used (if at all) are very simplistic and do not reflect all aspects of mobility. A high overall speed of nodes does not necessarily result in an increased probability of link breaks (e.g. if all the nodes move together with that high speed in the same direction). The only exception is again [85] which defines and uses a more sophisticated and reasonable mobility metric, already described in section 4.2.3.

4.5.3   Number of Nodes

In most scenarios the number of nodes is relatively low. Many scenarios just simulate up to 50 nodes, a few cases did simulate up to 400 nodes (e.g. in [21]). In my opinion a low number of nodes may be justified for certain kinds of scenarios, but it is certainly important to make more simulations with a higher number of nodes, possibly up to 10000.

Node density as an important factor has not been taken into account in some of the studies. Only in recent work does node density begin to receive more attention, which corresponds with efforts in the research community to develop more realistic mobility models (cf. [81]).

---

[6]There is no detailed description given in this book, due to the intended deadline. However, details can be obtained in the cited paper.

4.5.4 Modeling of Physical Properties

The *Smooth is Better than Sharp* mobility model[15] is an attempt to add physical constraints to the movement of nodes, i.e. direction changes cannot occur all of a sudden, they must be made by using turns. Further, the current speed has an impact on the turn radius. Speed changes are performed by acceleration and deceleration, and a direction change may also require a speed change first.

This is certainly important for more realistic scenarios, however, this model as proposed in [15] has not yet been used in simulations of ad hoc networks. The question is, whether such realistic modeling of physical movement constraints will have a noticeable impact on simulation results. I suspect, that this will not be the case, since such more realistic movements will not affect the density distribution within the area nor will they lead to a different number of link breaks.

The design of the *Smooth is Better than Sharp* model prevents a direct adaption in one of the common simulators, but it would be possible to modify it accordingly. So, it is very valuable as a reminder, to optionally add these physical constraints to future scenario generators. As NS-2 and GloMoSim do not support accelerated movement, and all discrete moves are straight, a turn and acceleration (and deceleration) must be emulated with intermediate steps. This will result in a tradeoff between accuracy and increased simulation time, due to the amount of intermediate steps.

4.5.5 Border Behavior

The use of the *Smooth is better than Sharp* model requires an explicit dealing with the crossing of a border. This is different to most scenarios which deal indirectly with the border (by not selecting "target" points beyond the area) or in a simple way (e.g. nodes are reflected).

Obviously, the problem of *border behavior* has not yet been handled in an appropriate and thorough manner. Either the problem is avoided or solved in a very simple way. It cannot be ignored though, since [15] and [29] have also shown that the border behavior has an important impact on the node distribution over the area, which affects density and therefore the simulation results.

A realistic approach would be to remove nodes crossing the area border. Then, it would be necessary to eventually introduce new nodes, which enter the area from another point on the border. Other possibilities include a "wraparound" border, that instantly transports the node to an opposite position, from where it will resume its movement[7](cf. [29] and [81]), or some kind of reflection method, as used in the *random direction* model.

Unfortunately, this aspect of the simulation depends very much on the capabilities of the simulation software. The removal and introduction of nodes during the simulated period is not yet supported in the commonly used simulation software packages.

4.6   Comparing Scenarios Against Benchmark Characteristics

In this section, the simulation scenarios are compared against a benchmark. The benchmark is not associated with any real application scenario. Instead of providing explicit values with which the scenarios are compared, the benchmark characteristics are defined more relaxed. The comparison functions yield the result according to a range of possible values. The reference values for the characteristics are chosen to eliminate unrealistic scenarios and to yield better results if a characteristic has a value within ranges, that might occur in practice and are thus applicable for real applications.

A detailed comparison against sample application scenarios is provided in chapter 5.

Like all comparisons, this comparison has been implemented using the *Perl* scripting language [162]. The program reads the data in a structured field representation, which is attached to the perl code, parses it and calls corresponding comparison functions for each characteristic.

The following section describes in detail, how the comparison against the benchmark characteristics is undertaken.

4.6.1   Benchmark Characteristics and Reference Values

The reference values for the characteristics used in this comparison define ranges of values which might occur in practice. It is assumed that such scenarios could use persons or cars as nodes and utilize radio equipment with a range of 200 - 700 meters.

The characteristics are not explicitly weighted for this comparison but contribute with the same proportion to the overall result.

Since reference values are fixed and directly integrated into the comparison functions, the functions only accept one parameter.

---

[7]This would result in an area shaped as a torus.

Thus, the overall result is computed as the average of the individual results:

$$f = \frac{1}{7}\sum_{i=1}^{7} f_i$$

Area Range

$$f_1 : (\mathbb{R}_0^+ \cup \{?\}) \mapsto [0 : 1]$$

$$f_1(d) = \begin{cases} 0.5 & \text{if } d =? \\ 1 & \text{if } 100 \le d \le 10000 \\ \frac{d}{100} & \text{if } d < 100 \\ e^{-(\frac{d-10000}{4000})} & \text{if } d > 10000 \end{cases}$$

The area range function compares the greatest distance d of the sim-ulation area against the reference value. A distance of 100 - 10000m yields the maximum value of 1.0. An ad hoc network within a smaller area would probably be of use only in certain special applications. A distance less than 100m results in the distance itself as result. A diam-eter larger than 10000m will yield a result according to the exponential function $e^{-(\frac{d-10000}{4000})}$.

This results in an exponentially smaller value the larger the excess distance. The parameters are chosen, such that with a distance of 10000 the result is still 1.0, at 20000 the result is close to 0.1, i.e. 0.082.

Number of Nodes The number of nodes n is compared against a reference interval with a range of 50 - 10000, which will yield 1.0. For more or less nodes, the result is compared in a similar fashion to the area range. Less than 50 nodes yield a result of 2· number of nodes, while more than 10000 nodes, yield a result according to the same formula as above:

$$f_2 : (\mathbb{R}_0^+ \cup \{?\}) \mapsto [0 : 1]$$

$$f_2(n) = \begin{cases} 0.5 & \text{if } n =? \\ 1 & \text{if } 50 \le n \le 10000 \\ \frac{n}{50} & \text{if } n < 50 \\ e^{-(\frac{n-10000}{4000})} & \text{if } n > 10000 \end{cases}$$

Node Density Since the node density δ is a very important parameter which significantly affects the performance of an ad hoc routing protocol, it is also compared against a reference interval. The density is computed by dividing the number of nodes by the size of the area and presented in nodes per square-meter. A value between 0.001 and 0.2 yields the maximum value. Values below 0.001 yield the density value $\cdot 1000$ and values above 0.2 yield $(1.2 - \delta)$. A very high density (larger than 1.0) should not yield a value better than 0.2. This assumes that the density does not exceed 1.2, which is true for all examined simulation scenarios.

$$f_3 : ([0 : 1.2] \cup \{?\}) \mapsto [0 : 1]$$

$$f_3(\delta) = \begin{cases} 0.5 & \text{if } \delta = ? \\ 1 & \text{if } 0.001 \leq \delta \leq 0.2 \\ 1000\delta & \text{if } \delta < 0.001 \\ 1.2 - \delta & \text{if } \delta > 0.2 \end{cases}$$

Obstacles Use of obstacles in the simulation scenario is specified by a boolean value (**Yes** or **No**).

For the use of movement obstacles m, as well as for radio obstacles r, 0.5 each are added, thus the maximum value is yielded only if both types of obstacles are used.

$$f_4 : \{Y, N, ?\} \times \{Y, N, ?\} \mapsto [0 : 1]$$

$f_4(r, m)$	Y	N	?
Y	1.0	0.5	0.75
N	0.5	0	0.25
?	0.75	0.25	0.5

Movement Strategy Depending on the strategy S a fixed value is returned: 0.5, if the strategy is unknown, 0.3 for $S = Random\ Direction$, 0.4 for $S = Random\ Waypoint$ and 1.0 for $S = Complex$.

$$f_5 : \{RandomDirection, RandomWaypoint, Complex, ?\} \mapsto [0 : 1]$$

$f_5(S)$	Random Direction	Random Waypoint	Complex	?
	0.3	0.4	1.0	0.5

Traffic Rate The traffic rate r is given in kbit/s and is computed from the number of traffic sources and their average rate. Only this overall traffic is measured against the benchmark values. If the rate is unknown then 0.5 are returned. For a rate between 2.4 and 2000 (imagine a 2400 baud modem up to 2Mbps WaveLAN in ad hoc mode) 1.0 is returned. Below 2.4 the value divided by 2.4 is returned and between 2000 and 54000 (there is 54Mbps WaveLAN now) the following value is returned: $(52000 - (rate - 2000))/52000$, a linear function that yields 1.0 at a rate of 2000 and 0 for 54000 and above.

$$f_6 : [2.4 : \infty] \mapsto [0 : 1]$$

$$f_6(r) = \begin{cases} 0.5 & \text{if} & r =? \\ r/2.4 & \text{if} & r < 2.4 \\ 1 & \text{if} & 2.4 \le r \le 2000 \\ (52000 - (r - 2000))/52000 & \text{if} & 2000 \le r \le 54000 \\ 0 & \text{else} \end{cases}$$

Traffic Type The traffic type τ can be UDP/constant bit rate or TCP/variable bit rate. UDP/constant bit rate traffic yields 0.3, TCP/variable bit rate traffic would yield 0.6, but was not used in the examined simulations. No other values are returned. The actual traffic type would probably be a mix, again depending on the application.

$$f_7 : \{UDP/CBR, TCP/VBR, ?\} \mapsto [0 : 1]$$

$f_5(S)$	UDP/CBR	TCP/VBR	?
	0.3	0.6	0.5

Example

To illustrate how the benchmark evaluation works, I selected the Johansson scenario "Disaster Area"[85], because it utilizes a lot of possible features and yields a very high value.

The values of its characteristics and the results according to the functions as described above are summarized in the following table:

Characteristic	Value	Result
Area Range	$d = \sqrt{1500^2 + 900^2} = \sqrt{3060000} \approx 1749$	1.0
Number of Nodes	$n = 50$	1.0
Node Density	$\delta = \frac{50}{1500 \cdot 900} = 1/27000$	0.037
Obstacles	r = "Y", m = "Y"	1.0
Movement Strategy	S = "Complex"	1.0
Traffic Rate	$r = 87 \cdot 20 = 1740$	1.0
Traffic Type	τ = "UDP/CBR"	0.3
Average		0.7624

4.6.2 Benchmark Results

The result of the evaluation is illustrated in figure 4.1. The maximum possible
value is 1.0, which means that each characteristic was within the ideal values
chosen in the evaluation.

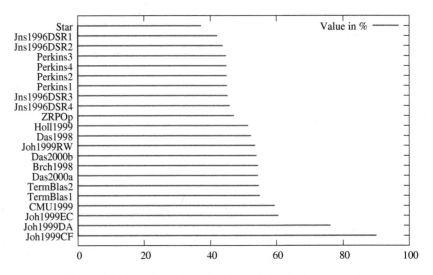

Figure 4.1: Benchmark evaluation of simulation scenarios

This very simple benchmark evaluation adds an additional, more rational argument to the criticism presented in section 4.5. The Johansson scenarios[85]
get the best results in this evaluation (conference room even gets 90%), all
other scenarios have a mediocre quality between 40% and 60%
The experiment for STAR [60] (as described in section 4.2.2) has the worst
value, since it is also very simple and in addition has a very low node density,

which may be useful only in very special applications and therefore yields poor values for the corresponding characteristics.

4.7 Comparison of the Simulation Scenarios

To establish a similarity relation for the previous simulation scenarios (and the experiments they were used for), I compare them according to the key characteristics of the simulation experiments. The result can be used to identify which experiments are similar enough to compare the results of the analysis. It can also be used to identify simulation environments which have been utilized in such experiments, and more important, which environments have never been used in an experiment.

A detailed description of each part of the comparison function is provided in the following section 4.7.1.

4.7.1 Characteristics used for Comparison

I present below the characteristics of evaluation experiments. Some of these characteristics have been discussed earlier in this chapter in section 4.1, but this section focuses on the individual functions used to compare these characteristics. If any value is unknown, this is denoted by a question mark (?). These characteristics are used to compare scenarios with themselves and partially to map them to applications.

Area The simulation area is assumed to be a rectangle. As mentioned above, all examined simulation scenarios do have a rectangular area. the length and width is specified in meters. For the comparison three different values are compared: The length, width and size of the area. I choose not to compare only the size of the area, since the shape can have a significant impact on the performance of routing in a MANET. Two areas of the same size, but one with a shape of a square, the other with a shape of a very stretched rectangle (length >> width or the other way round) are very different areas and should not yield the result of full match.

The length, width and size are each individually compared and the average of these results is used (added up and then divided by 3 since there are three values to be compared). The relative weight of this characteristic is 3 (in this case the normalizing factor and the weight outweigh each other, but this is just a special case):

$$f_1 : (\mathbb{R}_0^+ \cup \{?\})^6 \mapsto [0 : 1]$$

$$f_1(d_l, d_w, d_s, d_l', d_w', d_s') = \frac{1}{3}(g_1(d_l, d_l') + g_1(d_w, d_w') + g_1(d_s, d_s'))$$

with

$$g_1 : (\mathbb{R}_0^+ \cup \{?\}) \times (\mathbb{R}_0^+ \cup \{?\}) \mapsto [0 : 1]$$

$$g_1(x, x') = \begin{cases} 0.5 & \text{if} & x =? \wedge x' \neq ? \vee \\ & & x' =? \wedge x \neq ? \\ 0.25 & \text{if} & x =? \wedge x' =? \\ \frac{x}{x'} & \text{if} & \frac{x}{x'} < \frac{x'}{x} \\ \frac{x'}{x} & \text{else} \end{cases}$$

Node Type This compares the types of nodes, that are used in a scenario. More than one type can be specified. There are two node types, that appear in existing simulation scenarios: **Pedestrians** and **Cars**. Each scenario can specify a list of node types. These lists are compared and for each common node type, 1 is added to the intermediate result value, which is then divided by the maximum number of node types in a single scenario.

T and T' are sets of different node types occurring in scenario S and S'. Set X is the set of common nodetypes, which defined as follows:

$$X = \{x | x \in T \wedge x \in T' \wedge x \neq ?\}$$

u will be the additional factor relevant if there are unknown node types ("?"), which is defined as follows ($\mathbb{T}$ is the set of all possible type-sets T):

$$u : \mathbb{T} \times \mathbb{T} \mapsto \{0, 0.25, 0.5\}$$

$$u(T, T') = \begin{cases} 0.5 & \text{if} & ? \in T \wedge ? \notin T' \vee \\ & & ? \in T' \wedge ? \notin T \\ 0.25 & \text{if} & ? \in T \wedge ? \in T' \\ 0 & \text{else} \end{cases}$$

then the result is as follows:

$$f_2 : \mathbb{T} \times \mathbb{T} \mapsto [0 : 1]$$

$$f_2(T, T') = \frac{|X| + u(T, T')}{max(|T|, |T'|)}$$

Example: Scenario A with types {Pedestrian, Car} is compared to scenario B with only type {Car}, thus $T_A = \{P, C\}$, $T_B = \{C\}$, $X = \{C\}$ and the result is $\frac{|X|}{|T_A|} = \frac{1}{2}$. (One common node type divided by a maximum of two node types in scenario A).

The relative weight of this characteristic is 4.

Node speed The average speed of the nodes (regardless of type) is compared. Each scenario can specify a list of node speeds V which are compared to each other. The speed v is given in m/s (meter per second) in the experiments. V and V' are the ordered sets of occurring node speeds of the scenarios and $\mathbb{V}$ is the set of all possible sets V.

$$f_3 : \mathbb{V} \times \mathbb{V} \mapsto [0 : 1]$$

$$f_3(V, V') = \frac{\sum_{i=1}^{|V'|} g_3(v_i, v_i')}{max(|V|, |V'|)}$$

with

$$g_3 : (\mathbb{R}_0^+ \cup \{?\}) \times (\mathbb{R}_0^+ \cup \{?\}) \mapsto [0 : 1]$$

$$g_3(v, v') = \begin{cases} 0.5 & \text{if} & v = ? \wedge v' \neq ? \vee \\ & & v' = ? \wedge v \neq ? \\ 0.25 & \text{if} & v = ? \wedge v' = ? \\ \frac{v}{v'} & \text{if} & \frac{v}{v'} < \frac{v'}{v} \\ \frac{v'}{v} & \text{else} \end{cases}$$

The sum of the speed comparison values is divided by the maximum number of speed values specified for a single scenario.

The relative weight of this characteristic is 4.

Number of nodes The number of nodes n in the experiment is compared as follows:

$$f_4 : (\mathbb{N} \cup \{?\}) \times (\mathbb{N} \cup \{?\}) \mapsto [0:1]$$

$$f_4(n, n') = \begin{cases} 0.5 & \text{if} & n =? \wedge n' \neq ? \vee \\ & & n' =? \wedge n \neq ? \\ 0.25 & \text{if} & n =? \wedge n' =? \\ \frac{n}{n'} & \text{if} & \frac{n}{n'} < \frac{n'}{n} \\ \frac{n'}{n} & \text{else} \end{cases}$$

The relative weight of this characteristic is 5.

Duration of the experiment The duration of the simulation scenarios (in seconds) is compared as follows:

$$f_5 : (\mathbb{N} \cup \{?\}) \times (\mathbb{N} \cup \{?\}) \mapsto [0:1]$$

$$f_5(t, t') = \begin{cases} 0.5 & \text{if} & t =? \wedge t' \neq ? \vee \\ & & t' =? \wedge t \neq ? \\ 0.25 & \text{if} & t =? \wedge t' =? \\ \frac{t}{t'} & \text{if} & \frac{t}{t'} < \frac{t'}{t} \\ \frac{t'}{t} & \text{else} \end{cases}$$

The relative weight of this characteristic is 5.

Radio Range The range of the radio transmitters used in the simulation is again compared as follows:

$$f_6 : (\mathbb{R}_0^+ \cup \{?\}) \times (\mathbb{R}_0^+ \cup \{?\}) \mapsto [0:1]$$

$$f_6(r, r') = \begin{cases} 0.5 & \text{if} & r =? \wedge r' \neq ? \vee \\ & & r' =? \wedge r \neq ? \\ 0.25 & \text{if} & r =? \wedge r' =? \\ \frac{r}{r'} & \text{if} & \frac{r}{r'} < \frac{r'}{r} \\ \frac{r'}{r} & \text{else} \end{cases}$$

The relative weight of this characteristic is 2.

Use of Radio obstacles In theory there is a comparison function, which just compares whether the feature is used or not (indicated by α_{ro} equals **Y** or **N**), and yields a value of 0 or 1. In practice, there was no examined scenario which made use of obstacles, that influence the radio transmission (due to the fact, that common simulation software does not support such obstacles).

$$f_7 : \{Y, N, ?\} \times \{Y, N, ?\} \mapsto [0 : 1]$$

$f_7(\alpha_{ro}, \alpha'_{ro})$	Y	N	?
Y	1.0	0	0.5
N	0	1.0	0.5
?	0.5	0.5	0.25

The relative weight of this characteristic is 3.

Use of movement obstacles It is just compared whether movement obstacles are used or not (again indicated by α_{mo} equals **Y** or **N**). If there is a match, the comparison returns 1, a 0 is returned otherwise (except if there are unknown values of course).

$$f_8 : \{Y, N, ?\} \times \{Y, N, ?\} \mapsto [0 : 1]$$

$f_8(\alpha_{mo}, \alpha'_{mo})$	Y	N	?
Y	1.0	0	0.5
N	0	1.0	0.5
?	0.5	0.5	0.25

The relative weight of this characteristic is 3.

Use of Restricted Areas This is very similar to movement obstacles, (i.e. a large obstacle can be viewed as a restricted area). The difference is that a restricted area may only be restricted for a subset of nodes, or that some nodes may only move inside a certain area, while others may move more freely or are restricted to other areas.

It is compared whether such restricted areas are in use (α_{ra} equals **Y**) or not (α_{ra} equals **N**) just like the movement obstacles.

$$f_9 : \{Y, N, ?\} \times \{Y, N, ?\} \mapsto [0 : 1]$$

$f_9(\alpha_{ra}, \alpha'_{ra})$	Y	N	?
Y	1.0	0	0.5
N	0	1.0	0.5
?	0.5	0.5	0.25

The relative weight of this characteristic is 3.

Border Behaviour This compares the behaviour of the nodes β_{bb}, when they hit the border of the simulation area. Four cases are considered: **A**void running into the border (by the movement algorithm), **R**eflection of the nodes on border contact, i.e. the movement direction is changed, such that the node moves away from the border, **W**raparound, i.e. turning the area into a torus, a node which "leaves" the area by moving over one edge, reappears on the opposite border maintaining its movement direction angle and speed and **E**nter-Leave, a node which moves beyond a border has left the simulation and does not take part any more. Likewise new nodes are introduced by occasional random movements starting from a border and moving into the simulation area.

Apart from the obvious comparison functions, which return 1 for an exactly matching border behaviour, a different behaviour does not necessarily result in a value of 0. Some behaviours are more related than others, and this is reflected by some additional rules:

- *Avoid* and *Reflect* behaviour is defined to be similar by $0, 5$.

- *Wraparound* is considered to be similar by 0.2 compared to *Avoid* and *Reflect*, and similar by 0.3 compared to *Enter/Leave*.

- Other combinations are not considered similar (i.e. 0).

$$f_{10} : \{A, R, W, E, ?\} \times \{A, R, W, E, ?\} \mapsto [0 : 1]$$

$f_{10}(\beta_{bb}, \beta'_{bb})$	A	R	W	E	?
A	1.0	0.5	0.2	0	0.5
R	0.5	1.0	0.2	0	0.5
W	0.2	0.2	1.0	0.3	0.5
E	0	0	0.3	1.0	0.5
?	0.5	0.5	0.5	0.5	0.25

The relative weight of this characteristic is 3.

Introduction of New Nodes This compares, if in any scenario allows the introduction of new nodes during the simulation (indicated by α_{nn} equals **Y** or **N**). The same rules as for movement obstacles apply.

$$f_{11} : \{Y, N, ?\} \times \{Y, N, ?\} \mapsto [0 : 1]$$

$f_{11}(\alpha_{nn}, \alpha'_{nn})$	Y	N	?
Y	1.0	0	0.5
N	0	1.0	0.5
?	0.5	0.5	0.25

The relative weight of this characteristic is 3.

Removal of Nodes Likewise it is compared if nodes can also be removed during the simulation (indicated by α_{rn} equals **Y** or **N**).

$$f_{12} : \{Y, N, ?\} \times \{Y, N, ?\} \mapsto [0 : 1]$$

$f_{12}(\alpha_{rn}, \alpha'_{rn})$	Y	N	?
Y	1.0	0	0.5
N	0	1.0	0.5
?	0.5	0.5	0.25

The relative weight of this characteristic is 3.

Group Mobility This compares if group-mobility has been used for the experiment. Again it is only compared if group mobility is used or not (indicated by α_{gm} equals **Y** or **N**).

$$f_{13} : \{Y, N, ?\} \times \{Y, N, ?\} \mapsto [0 : 1]$$

$f_{13}(\alpha_{gm}, \alpha'_{gm})$	Y	N	?
Y	1.0	0	0.5
N	0	1.0	0.5
?	0.5	0.5	0.25

The relative weight of this characteristic is 3.

Movement Strategy This compares the basic movement strategy used in the simultion. The following movement strategies are defined: **RW** (*Random Waypoint*) (cf. section 4.2.2), **RD** (*Random Direction*) (cf. section 4.2.2), **B** (*Brownian Movement*) and **C** (*Complex Strategy*), which represents anything more sophisticated, than the other strategies.

Again here are some additional rules applied:

- *Random Waypoint* and *Random Direction* are considered to be similar by 0.5 and
- *Random Waypoint* and *Complex Strategy* are considered to be similar by 0.2.

$$f_{14} : \{RW, RD, B, C, ?\} \times \{RW, RD, B, C, ?\} \mapsto [0 : 1]$$

$f_{14}(\beta_{ms}, \beta'_{ms})$	RW	RD	B	C	?
RW	1.0	0.5	0	0.2	0.5
RD	0.5	1.0	0	0	0.5
B	0	0	1.0	0	0.5
C	0.2	0	0	1.0	0.5
?	0.5	0.5	0.5	0.5	0.25

The relative weight of this characteristic is 4.

Type of Traffic The type of traffic used in the simulation is compared, this is only a very rough characteristic. There are two subtypes of traffic. The first subtype compares the transport protocol used in the simulation valid values are **TCP** and **UDP**. The second subtype is a little bit application related and states if the traffic is of a constant bit rate or a variable bit rate (**CBR** and **VBR**).

Both are compared, but the bitrate subtype (β_{ttr}) contributes $\frac{3}{4}$, while the protocol subtype (β_{ttp}) only contributes $\frac{1}{4}$ to the result.

However, all experiments used **CBR-UDP** traffic.

$$\mathbb{D}_{ttp} = \{TCP, UDP, ?\}$$
$$\mathbb{D}_{ttr} = \{CBR, VBR, ?\}$$

$$f_{15} : \mathbb{D}_{ttp}^2 \times \mathbb{D}_{ttr}^2 \mapsto [0 : 1]$$

$$f_{15}(\beta_{ttp}, \beta'_{ttp}, \beta_{ttr}, \beta'_{ttr}) = \frac{1}{4}(g_{15}(\beta_{ttp}, \beta'_{ttp}) + 3g_{15}(\beta_{ttr}, \beta'_{ttr}))$$

$$g_{15} : (\mathbb{D}_{ttp} \cup \mathbb{D}_{ttr})^2 \mapsto [0 : 1]$$

$$g_{15}(x, x') = \begin{cases} 1.0 & \text{if} & x = x' \neq ? \\ 0.5 & \text{if} & x = ? \wedge x' \neq ? \vee \\ & & x' = ? \wedge x \neq ? \\ 0.25 & \text{if} & x = x' = ? \\ 0 & \text{else} \end{cases}$$

The overall weight of this comparison is 3.

Traffic Rate The rate of generated traffic is also compared. Two values are examined, the number of traffic sources (t_{ts}) and the individual rate of each source (t_{tr}) in kBit/s. A third value is computed, the overall traffic rate, which is just the product $t_{tr} \cdot t_{ts}$. First each scenario specifies the number of traffic sources Each of these values is compared separately according to the function g_{16} and the average value is computed as given below.

$$f_{16} : (\mathbb{R}_0^+ \cup \{?\})^4 \mapsto [0:1]$$

$$f_{16}(r_{ts}, r_{tr}, r'_{ts}, r'_{tr}) = \frac{1}{3}(g_{16}(r_{ts}, r'_{ts}) + g_{16}(r_{tr}, r'_{tr}) + g_{16}(r_{ts} \cdot r_{tr}, r'_{ts} \cdot r'_{tr}))$$

$$g_{16}(x, x') = \begin{cases} 0.5 & \text{if} & x =? \wedge x' \neq ?\vee \\ & & x' =? \wedge x \neq ? \\ 0.25 & \text{if} & x = x' =? \\ \frac{x}{x'} & \text{if} & \frac{x}{x'} < \frac{x'}{x} \\ \frac{x'}{x} & \text{else} \end{cases}$$

The overall weight of this comparison is 3.

As described, each individual function has a weight associated with it. This initial setting of weights has been chosen to give nearly equal weight to most of the functions with some emphasis on node type and speed, number of nodes and duration, as well as movement strategy, since it is my understanding, that these characteristics are those with more relevance to model a real application.

Considering the weights of each partial function, the final result is then computed as follows:

$f = \frac{1}{54}(3f_1 + 4f_2 + 4f_3 + 5f_4 + 5f_5 + 2f_6 + 3f_7 + 3f_8 + 3f_9 + 3f_{10} + 3f_{11} + 3f_{12} + 3f_{13} + 4f_{14} + 3f_{15} + 3f_{16})$

Example

In order to illustrate how the comparison works, the following example computes the comparison results for the "Broch, Maltz, Johnson, Hu and Jetcheva, 1998" [25] with the "Johansson Disaster Area" [85] scenario.

Characteristic	Value Broch1998	Value Johansson	W	Result
Area	$d_l = 1500$	$d'_l = 1500$	3	1.67
	$d_w = 300$	$d'_w = 900$		
	$d_s = 450000$	$d'_s = 1350000$		
Node Type	$T = \{\text{Person}, \text{Car}\}$	$T' = \{\text{Person}, \text{Car}\}$	4	4.0
Node Speed	$V = \{1, 20\}$	$V' = \{1, 20\}$	4	4.0
Number of nodes	$n = 50$	$n' = 50$	5	5.0
Duration	$t = 900$	$t' = 900$	5	5.0
Radio Range	$r = 250$	$r' = 250$	2	2.0
Radio Obstacles	$\alpha_{ro} = \text{``}N\text{''}$	$\alpha'_{ro} = \text{``}Y\text{''}$	3	0.0
Movement Obstacles	$\alpha_{mo} = \text{``}N\text{''}$	$\alpha'_{mo} = \text{``}Y\text{''}$	3	0.0
Restricted Areas	$\alpha_{ra} = \text{``}N\text{''}$	$\alpha'_{ra} = \text{``}Y\text{''}$	3	0.0
Border Behaviour	$\beta_{bb} = \text{``}Avoid\text{''}$	$\beta'_{bb} = \text{``}Avoid\text{''}$	3	3.0
Introduction of	$\alpha_{nn} = \text{``}N\text{''}$	$\alpha'_{nn} = \text{``}N\text{''}$	3	3.0
new nodes				
Removal of nodes	$\alpha_{rn} = \text{``}N\text{''}$	$\alpha'_{rn} = \text{``}N\text{''}$	3	3.0
Group Mobility	$\alpha_{gm} = \text{``}N\text{''}$	$\alpha'_{gm} = \text{``}Y\text{''}$	3	0.0
Movement Strategy	$\beta_{bb} = \text{``}RW\text{''}$	$\beta'_{bb} = \text{``}CX\text{''}$	4	0.8
Traffic Type	$\beta_{ttp} = \text{``}UDP\text{''}$	$\beta'_{ttp} = \text{``}UDP\text{''}$	3	3.0
	$\beta_{ttr} = \text{``}CBR\text{''}$	$\beta'_{ttr} = \text{``}CBR\text{''}$		
Traffic Rate	$r_{ts} = 30$	$r'_{ts} = 87$	3	1.56
	$r_{tr} = 26$	$r'_{tr} = 20$		
Sum:			54	36.02
Result				0.67

4.7.2 Characterizing and Comparing the Scenarios

I will now characterize each simulation scenario used in the later comparison in chapter 5. Each description will include a *Label*, which denotes the experiment (which applied the simulation scenario). The attached diagrams display the degree of similarity of each scenario to each other.

Johnson and Maltz, 1996

This set of experiments was described in [86]. The experiments differ only in the number of nodes used and the movement speed.

Labels: Jns1996DSR1, Jns1996DSR2,
                                 Jns1996DSR3, Jns1996DSR4
Area: $9x9m$
Type of nodes: Person
Node speed: $0.3 - 0.7m/s$
Number of Nodes: 6,12,18 and 24
Duration: $4000s$
Transmission Range: $3m$
Radio Obstacles: No
Movement Obstacles: No
Restricted Areas: No
Border Behavior: Avoid
Introduction of new nodes: No
Removal of nodes: No
Group Mobility: No
Strategy: Random Waypoint
Type of Traffic: CBR, UDP
Traffic-rate: 18*35 kbit/s

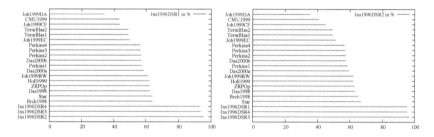

Figure 4.2: Similarities of the Johnson and Maltz experiments 1 and 2

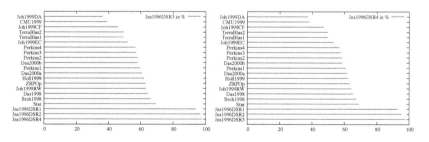

Figure 4.3: Similarities of the Johnson and Maltz experiments 3 and 4

As expected all four scenarios of these experiments match each other very well. The Johansson scenarios (disaster area) do match least. These experiments were done on a very small scale, so most experiments which used larger areas do not match a lot, except the Star experiment. This also used a very low number of nodes and is also very simple, thus is yields a good match although the difference in the area size could not be bigger.

Broch, Maltz, Johnson, Hu and Jetcheva, 1998

This experiment compared four routing protocols and is found in [25].

Label:	Brch1998
Area:	$1500x300m$
Type of nodes:	Person, Car
Node speed:	$1, 20m/s$
Number of Nodes:	50
Duration:	$900s$
Transmission Range:	$250m$
Radio Obstacles:	No
Movement Obstacles:	No
Restricted Areas:	No
Border Behavior:	Avoid
Introduction of new nodes:	No
Removal of nodes:	No
Group Mobility:	No
Strategy:	Random Waypoint
Type of Traffic:	CBR, UDP
Traffic-rate:	10,20,30*26 kbit/s

This experiment uses rather common values for the area, number of nodes, Strategy and duration. Thus it matches fairly to most other scenarios. The Das2000a experiment matches best, since it nearly matches all characteristics exacly. The only differences are that it only uses cars but no persons and the traffic rate is slightly different.

Das, Castañeda, Yan and Sengupta, 1998

This experiment compared multiple protocols and is described in [44].

Label:	Das1998
Area:	$1000x1000m$
Type of nodes:	Person
Node speed:	$0.6, 4.5m/s$
Number of Nodes:	60
Duration:	$10000s$
Transmission Range:	$350m$
Radio Obstacles:	No
Movement Obstacles:	No
Restricted Areas:	No
Border Behavior:	Reflect
Introduction of new nodes:	No
Removal of nodes:	No
Group Mobility:	No
Strategy:	Random Direction
Type of Traffic:	CBR, UDP
Traffic-rate:	60*15 kbit/s

This experiment matches well with the smaller Perkins experiments and with the ZRP experiment by Haas and Pearlman. They share the same strategy

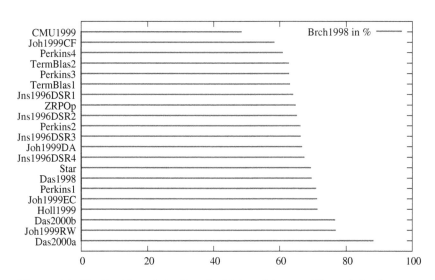

Figure 4.4: Similarities of the Broch, Maltz, Johnson, Hu and Jetcheva experiments

similar numbers of nodes, area size and transmission range.

Holland and Vaidya, 1999

This experiment deals with TCP performance and is described in [73].

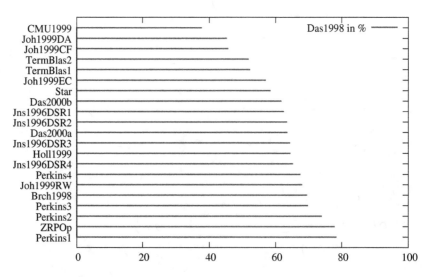

Figure 4.5: Similarities of the Das, Castañeda, Yan and Sengupta experiments

Label:	Holl1999
Area:	$1500x300m$
Type of nodes:	?
Node speed:	? m/s
Number of Nodes:	?
Duration:	?
Transmission Range:	$250m$
Radio Obstacles:	No
Movement Obstacles:	No
Restricted Areas:	No
Border Behavior:	Avoid
Introduction of new nodes:	No
Removal of nodes:	No
Group Mobility:	No
Strategy:	?
Type of Traffic:	TCP
Traffic-rate:	?*? kbit/s

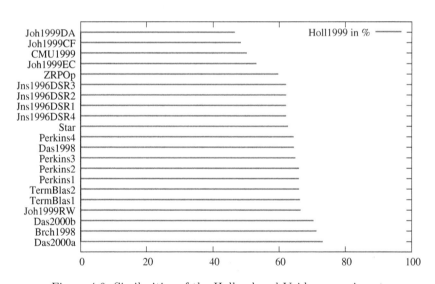

Figure 4.6: Similarities of the Holland and Vaidya experiments

Similar as Broch1998 (cf. 4.7.2) very common values are used for the characteristics and thus a rather high degree of similarity is achieved to most other experiments.

DSR Testbed

This real testbed experiment was done by Maltz, Broch and Johnson 1999, cf. [111].

Label:	CMU1999
Area:	$1000x300m$
Type of nodes:	Car
Node speed:	20 m/s
Number of Nodes:	7
Duration:	?
Transmission Range:	$250m$
Radio Obstacles:	Yes
Movement Obstacles:	Yes
Restricted Areas:	Yes
Border Behavior:	Avoid
Introduction of new nodes:	No
Removal of nodes:	No
Group Mobility:	No
Strategy:	Complex
Type of Traffic:	?,?
Traffic-rate:	?*? kbit/s

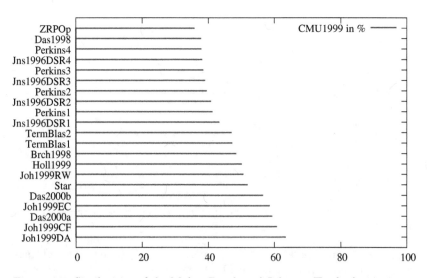

Figure 4.7: Similarities of the Maltz, Broch and Johnson Testbed experiment

This scenario has a very low number of nodes , apart from that it uses obstacles (since it was a real world experiment and no simulation). Interestingly it matches best with the Johansson scenarios which also try to take realistic node behaviour and obstacles into account. However, the best match is just over 60%, while many are even below 40%. This experiment is obviously very different from the majority, which is also confirmed by the degree of similarity to the other experiments.

The Johansson Experiments

The following experiments are the notable achievement of Per Johansson et. al. during a serious attempt to add more realism to evaluations of mobile ad hoc network Routing Protocols. They are all described in [85].

Label:	Joh1999CF
Area:	150x90
Type of nodes:	Persons
Node speed:	1 m/s
Number of Nodes:	50
Duration:	900
Transmission Range:	25m
Radio Obstacles:	Yes
Movement Obstacles:	Yes
Restricted Areas:	Yes
Border Behavior:	Avoid
Introduction of new nodes:	No
Removal of nodes:	No
Group Mobility:	No
Strategy:	Complex
Type of Traffic:	CBR,UDP
Traffic-rate:	6*20 kbit/s

Label:	Joh1999RW
Area:	$1000x1000$
Type of nodes:	?
Node speed:	$? \ m/s$
Number of Nodes:	50
Duration:	250
Transmission Range:	$250m$
Radio Obstacles:	No
Movement Obstacles:	No
Restricted Areas:	No
Border Behavior:	Avoid
Introduction of new nodes:	No
Removal of nodes:	No
Group Mobility:	No
Strategy:	Random Waypoint
Type of Traffic:	CBR,UDP
Traffic-rate:	15*3.2 kbit/s

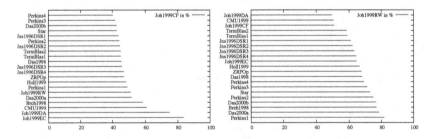

Figure 4.8: Similarities of the Johansson experiments CF and RW

Label:	Joh1999EC
Area:	1500x900
Type of nodes:	Person
Node speed:	1 m/s
Number of Nodes:	50
Duration:	900
Transmission Range:	250m
Radio Obstacles:	Yes
Movement Obstacles:	Yes
Restricted Areas:	Yes
Border Behavior:	Avoid
Introduction of new nodes:	No
Removal of nodes:	No
Group Mobility:	No
Strategy:	Random Waypoint
Type of Traffic:	CBR,UDP
Traffic-rate:	?*20 kbit/s

Label:	Joh1999DA
Area:	1500x900
Type of nodes:	Persons, Cars
Node speed:	1,20 m/s
Number of Nodes:	50
Duration:	900
Transmission Range:	250m
Radio Obstacles:	Yes
Movement Obstacles:	Yes
Restricted Areas:	Yes
Border Behavior:	Avoid
Introduction of new nodes:	No
Removal of nodes:	No
Group Mobility:	Yes
Strategy:	Complex
Type of Traffic:	CBR,UDP
Traffic-rate:	87*20 kbit/s

The Johansson experiments match each other rather well (as expected), except for the restricted random waypoint scenario, which was designed different on purpose by Johansson et al.. Most other (more simple) scenarios do not match them very well, except Broch1998 and also the CMU testbed.

Das, Perkins and Royer, 2000

These experiments compare AODV and DSR and are described in [45].

Label:	Das2000a
Area:	$1500x300$
Type of nodes:	Cars
Node speed:	$20\ m/s$
Number of Nodes:	50
Duration:	900
Transmission Range:	$250m$
Radio Obstacles:	No
Movement Obstacles:	No
Restricted Areas:	No
Border Behavior:	Avoid
Introduction of new nodes:	No
Removal of nodes:	No
Group Mobility:	No
Strategy:	Random Waypoint
Type of Traffic:	CBR,UDP
Traffic-rate:	10,20,30,40*20 kbit/s

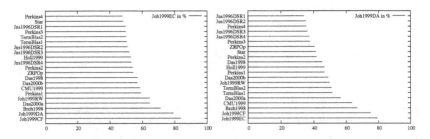

Figure 4.9: Similarities of the Johansson experiments EC and DA

Label:	Das2000b
Area:	$2200x600$
Type of nodes:	Cars
Node speed:	20 m/s
Number of Nodes:	100
Duration:	500
Transmission Range:	$250m$
Radio Obstacles:	No
Movement Obstacles:	No
Restricted Areas:	No
Border Behavior:	Avoid
Introduction of new nodes:	No
Removal of nodes:	No
Group Mobility:	No
Strategy:	Random Waypoint
Type of Traffic:	CBR,UDP
Traffic-rate:	10,20,30,40*20 kbit/s

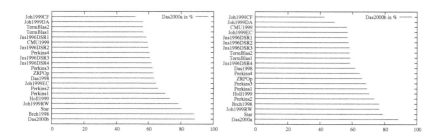

Figure 4.10: Similarities of the Das, Perkins and Royer experiments

Again two scenarios, which use common values for most paramters, especially
the first experiment. Consequently it matches may other experiments rather
well. The second one has a larger area and thus has its best matches (apart
from the sibling scenario) in the set that use large areas and low node density.

Blazević, Giordano and Le Boudec, 2000

These experiments deal with Terminode routing (a wide area ad hoc networking mechanism) and are described in [21].

Label:	TermBlas1
Area:	$3500x2500$
Type of nodes:	Persons,Cars
Node speed:	10,20 m/s
Number of Nodes:	400
Duration:	?
Transmission Range:	$250m$
Radio Obstacles:	No
Movement Obstacles:	No
Restricted Areas:	Yes
Border Behavior:	Avoid
Introduction of new nodes:	No
Removal of nodes:	No
Group Mobility:	No
Strategy:	Random Waypoint
Type of Traffic:	CBR,UDP
Traffic-rate:	30*2 kbit/s

Label:	TermBlas2
Area:	$4500x3500$
Type of nodes:	Persons,Cars
Node speed:	10,20 m/s
Number of Nodes:	400
Duration:	?
Transmission Range:	$250m$
Radio Obstacles:	No
Movement Obstacles:	No
Restricted Areas:	Yes
Border Behavior:	Avoid
Introduction of new nodes:	No
Removal of nodes:	No
Group Mobility:	No
Strategy:	Random Waypoint
Type of Traffic:	CBR,UDP
Traffic-rate:	30*2 kbit/s

These scenarios model the use of wide area ad hoc networking as intended by the terminode project [154]. Apart from each other, they match other scenarios which used large areas and great distances like Perkins4.

Perkins and Royer, 2001

These experiments cover AODV and are described in [128]. GloMoSim has been used in these experiments. There have been just performance evaluations in terms of routing overhead and delivery ration, no comparisons with other routing protocols.

Labels:	Perkins1, Perkins2, Perkins3, Perkins4
Area:	$1000x1000, 1500x1500, 2400x2400, 3450x3450$
Type of nodes:	Persons
Node speed:	$10\ m/s$
Number of Nodes:	50,100,200,500
Duration:	300
Transmission Range:	$250m$
Radio Obstacles:	No
Movement Obstacles:	No
Restricted Areas:	No
Border Behavior:	Reflect
Introduction of new nodes:	No
Removal of nodes:	No
Group Mobility:	No
Strategy:	Random Direction
Type of Traffic:	CBR,UDP
Traffic-rate:	?*? kbit/s

As expected these experiments match each other best, then they match best with other experiments with large areas, and with a high number of nodes. Among the next best matches is the Johansson restricted random waypoint scenario and the ZRP experiment, which also shares the same strategy (ran-

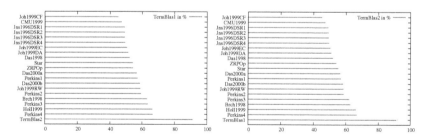

Figure 4.11: Similarities of the Blazević, Giordano and Le Boudec experiments

dom direction).

Haas and Pearlman, 1999

This experiments evaluates ZRP using the OPNET simulator. It is described in [69].

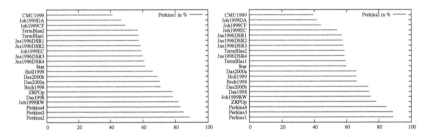

Figure 4.12: Similarities of the Perkins and Royer experiments 1 and 2

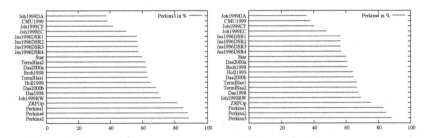

Figure 4.13: Similarities of the Perkins and Royer experiments 3 and 4

Label:	ZRPop
Area:	$1000x1000$
Type of nodes:	Persons
Node speed:	$?\ m/s$
Number of Nodes:	200
Duration:	?
Transmission Range:	$105m$
Radio Obstacles:	No
Movement Obstacles:	No
Restricted Areas:	No
Border Behavior:	Reflect
Introduction of new nodes:	No
Removal of nodes:	No
Group Mobility:	No
Strategy:	Random Direction
Type of Traffic:	CBR,UDP
Traffic-rate:	?*? kbit/s

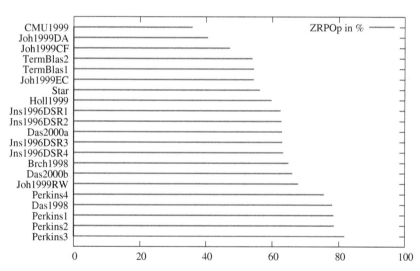

Figure 4.14: Similarities of the Haas and Pearlman experiments

This experiment uses a rather high number of nodes, a larger than average are size and uses random direction thus it matches very well with the Perkins experiments and with the Das1998 experiment.

Garcia-Luna-Aceves and Spohn, 2001

This experiment evaluates STAR and is described in [60]. It was done using a *C++ Protocol Toolkit* and compared STAR with a traditional link state approaches and a method called ALP[8]. The experiment measured packet overhead, connectivity changes and delivery ratios.

Label:	Star
Area:	$5000x7000$
Type of nodes:	Cars
Node speed:	$20\ m/s$
Number of Nodes:	20
Duration:	?
Transmission Range:	$105m$
Radio Obstacles:	No
Movement Obstacles:	No
Restricted Areas:	No
Border Behavior:	Reflect
Introduction of new nodes:	No
Removal of nodes:	No
Group Mobility:	No
Strategy:	Random Waypoint
Type of Traffic:	CBR,UDP
Traffic-rate:	?*? kbit/s

This experiment uses a very large area, but only a very small number of nodes. It matches best the Das2000 experiments, which also use cars with $20m/s$.

4.7.3   Similarity Results

As described above, the experiments have been compared according to their characteristics. The following diagram illustrates the level of similarity of each experiment to all other experiments. The graphs that shows the relations of all experiments which have a degree of similarity of at least 60% and 70% are shown in figures 4.16 and 4.17 . The colors of the edges denote the degree of similarity[9]:

---

[8]probably "adaptive link state protocol", but wrongly described as "Account, Login, Password"

[9]Please note that in the tables the colors are used as well to highlight the range of the value. However, due to rounding, some border values (e.g. 70%) may appear in different colors.

Color	Similarity s in %
red	$s \geq 90$
orange	$80 \leq s < 90$
light green	$70 \leq s < 80$
dark green	$60 \leq s < 70$

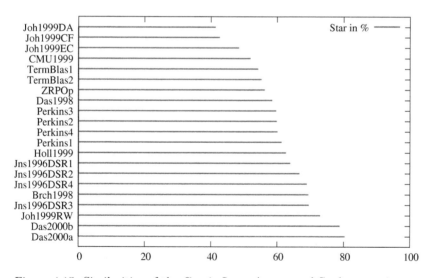

Figure 4.15: Similarities of the Garcia-Luna-Aceves and Spohn experiments

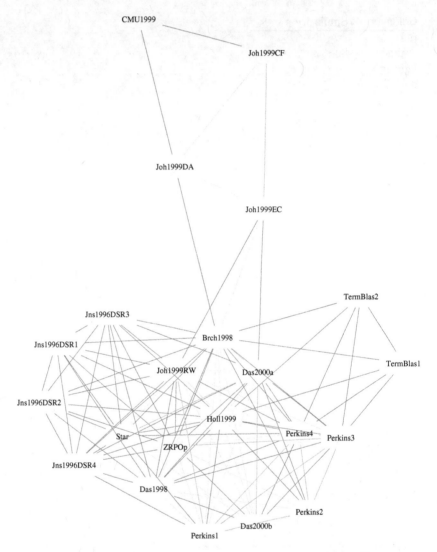

Figure 4.16: Similarities of previous experiments $\geq 60\%$

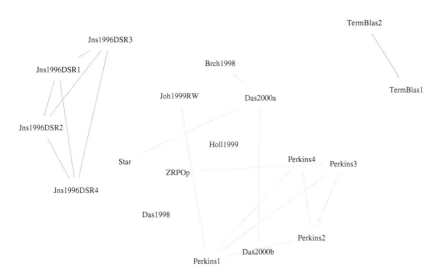

Figure 4.17: Similarities of previous experiments $\geq 70\%$

Proto	Brch1998	CMU1999	Das1998	Das2000a	Das2000b
Brch1998	–	0.48	0.70	0.88	0.77
CMU1999	0.48	–	0.38	0.59	0.57
Das1998	0.70	0.38	–	0.64	0.62
Das2000a	0.88	0.59	0.64	–	0.89
Das2000b	0.77	0.57	0.62	0.89	–
Holl1999	0.71	0.50	0.65	0.73	0.70
Jns1996DSR1	0.64	0.43	0.63	0.59	0.57
Jns1996DSR2	0.65	0.41	0.63	0.60	0.58
Jns1996DSR3	0.66	0.39	0.64	0.61	0.58
Jns1996DSR4	0.67	0.38	0.65	0.62	0.59
Joh1999CF	0.58	0.61	0.46	0.52	0.43
Joh1999DA	0.67	0.63	0.45	0.56	0.49
Joh1999EC	0.71	0.59	0.57	0.65	0.57
Joh1999RW	0.72	0.51	0.68	0.70	0.67
Perkins1	0.71	0.41	0.75	0.71	0.69
Perkins2	0.66	0.40	0.71	0.66	0.73
Perkins3	0.63	0.38	0.70	0.62	0.68
Perkins4	0.61	0.38	0.68	0.61	0.65
Star	0.69	0.52	0.59	0.80	0.79
TermBlas1	0.63	0.47	0.52	0.57	0.59
TermBlas2	0.63	0.47	0.52	0.56	0.58
ZRPOp	0.65	0.36	0.78	0.63	0.66

Proto	Holl1999	Jns1996DSR1	Jns1996DSR2	Jns1996DSR3	Jns1996DSR4
Brch1998		0.64	0.65	0.66	0.67
CMU1999	0.50	0.43	0.41	0.39	0.38
Das1998	0.65	0.63	0.63	0.64	0.65
Das2000a		0.59	0.60	0.61	0.62
Das2000b		0.57	0.58	0.58	0.59
Holl1999	–	0.62	0.62	0.62	0.62
Jns1996DSR1	0.62	–	0.95	0.94	0.93
Jns1996DSR2	0.62	0.95	–	0.97	0.95
Jns1996DSR3	0.62	0.94	0.97	–	0.98
Jns1996DSR4	0.62	0.93	0.95	0.98	–
Joh1999CF	0.48	0.44	0.45	0.46	0.47
Joh1999DA	0.47	0.34	0.35	0.36	0.37
Joh1999EC	0.53	0.50	0.51	0.52	0.53
Joh1999RW	0.67	0.61	0.62	0.63	0.64
Perkins1	0.66	0.58	0.59	0.60	0.61
Perkins2	0.66	0.57	0.58	0.58	0.59
Perkins3	0.65	0.57	0.57	0.57	0.58
Perkins4	0.64	0.57	0.57	0.57	0.57
Star	0.63	0.64	0.67	0.70	0.69
TermBlas1	0.66	0.49	0.49	0.49	0.49
TermBlas2	0.66	0.49	0.49	0.49	0.49
ZRPOp	0.60	0.62	0.63	0.63	0.63

Proto	Joh1999CF	Joh1999DA	Joh1999EC	Joh1999RW	Perkins1
Brch1998	0.58	0.67	0.74	0.77	0.71
CMU1999	0.61	0.63	0.59	0.51	0.41
Das1998	0.46	0.45	0.57	0.68	0.78
Das2000a	0.52	0.56	0.65	0.79	0.71
Das2000b	0.43	0.49	0.57	0.77	0.69
Holl1999	0.48	0.47	0.53	0.67	0.66
Jns1996DSR1	0.44	0.34	0.50	0.61	0.58
Jns1996DSR2	0.45	0.35	0.51	0.62	0.59
Jns1996DSR3	0.46	0.36	0.52	0.63	0.60
Jns1996DSR4	0.47	0.37	0.53	0.64	0.61
Joh1999CF	–	0.75	0.84	0.51	0.49
Joh1999DA	0.75	–	0.79	0.50	0.47
Joh1999EC	0.84	0.79	–	0.65	0.59
Joh1999RW	0.51	0.50	0.65	–	0.82
Perkins1	0.49	0.47	0.59	0.82	–
Perkins2	0.44	0.42	0.54	0.75	0.89
Perkins3	0.42	0.38	0.50	0.71	0.85
Perkins4	0.40	0.36	0.48	0.69	0.83
Star	0.43	0.41	0.48	0.71	0.61
TermBlas1	0.45	0.51	0.51	0.59	0.57
TermBlas2	0.45	0.51	0.50	0.59	0.57
ZRPOp	0.47	0.41	0.54	0.68	0.78

Proto	Perkins2	Perkins3	Perkins4	Star	TermBlas1
Brch1998	0.66	0.63	0.61	0.69	0.63
CMU1999	0.40	0.38	0.38	0.52	0.47
Das1998		0.70	0.68	0.59	0.52
Das2000a	0.66	0.62	0.61	0.80	0.57
Das2000b		0.68	0.65		0.59
Holl1999	0.66	0.65	0.64	0.63	0.66
Jns1996DSR1	0.57	0.57	0.57	0.64	0.49
Jns1996DSR2	0.58	0.57	0.57	0.67	0.49
Jns1996DSR3	0.58	0.57	0.57	0.70	0.49
Jns1996DSR4	0.59	0.58	0.57	0.69	0.49
Joh1999CF	0.44	0.42	0.40	0.43	0.45
Joh1999DA	0.42	0.38	0.36	0.41	0.51
Joh1999EC	0.54	0.50	0.48	0.48	0.51
Joh1999RW			0.69		0.59
Perkins1	0.89	0.85	0.83	0.61	0.57
Perkins2	–	0.89	0.85	0.60	0.59
Perkins3	0.89	–	0.88	0.60	0.64
Perkins4	0.85	0.88	–	0.60	0.67
Star	0.60	0.60	0.60	–	0.54
TermBlas1	0.59	0.64	0.67	0.54	–
TermBlas2	0.59	0.62	0.67	0.55	0.91
ZRPOp		0.82		0.56	0.54

Proto	TermBlas2	ZRPOp
Brch1998	0.63	0.65
CMU1999	0.47	0.36
Das1998	0.52	0.78
Das2000a	0.56	0.63
Das2000b	0.58	0.66
Holl1999	0.66	0.60
Jns1996DSR1	0.49	0.62
Jns1996DSR2	0.49	0.63
Jns1996DSR3	0.49	0.63
Jns1996DSR4	0.49	0.63
Joh1999CF	0.45	0.47
Joh1999DA	0.51	0.41
Joh1999EC	0.50	0.54
Joh1999RW	0.59	0.68
Perkins1	0.57	0.78
Perkins2	0.59	0.78
Perkins3	0.62	0.82
Perkins4	0.67	0.77
Star	0.55	0.56
TermBlas1	0.91	0.54
TermBlas2	–	0.54
ZRPOp	0.54	–

These diagrams illustrate that most previous experiments used a similar environment. Except a few instances stand out, these are the "Terminode" experiments[21], the Johansson experiments [85] (without the restricted random waypoint, which is more similar to the majority of experiments than to the other Johansson expreiments), the CMU testbed[111] experiment and the small-scale Johnson experiments. Especially the fact that the experiments most similar to the real world (CMU testbed[111]) are the Johansson experiments, supports the suspicion that more experiments with this kind of realistic environment are required and "simple scenarios" may not be good enough. However, it remains to be determined if the results of routing analysis done with simple and realistic scenarios will differ significantly.

Chapter 5

Applications for MANETs

This chapter introduces a set of application scenarios that could be used with mobile ad hoc networking. To design a scenario for the evaluation of a routing protocol for mobile ad hoc networks the possible applications need to be taken into account. The requirements for a simulation scenario then need to be derived from the intended application.

5.1 Overview

The following applications are proposed to be the main fields where mobile ad hoc networks could be used. However, after a closer look, not all of them seem to be appropriate as a sample scenario for mobile ad hoc networks.

1. Personal Area Networks (PAN), Bluetooth Scatter Networks

2. Conference Room networks

3. Trade Fair networks

4. Extended Cellular Phone Networks

5. Office Building Networks

6. Individual Spontaneous Networks (People in a neighborhood, town, city or even larger scale, form such a network)

7. Car Based networks

8. Farm and Park Management

9. Sensor networks

10. Disaster Area Recovery Support

11. Military Applications

For each of these applications, I have attached a diagram, that shows how good the evaluation scenarios (as described in chapter 4) match this application. The match is very rough and basic and can only have comparative value. This will be explained more detailed in section 5.3.

5.2   Characteristics of Application Scenarios

I analyse these sample applications according to some key characteristics and requirements. I will use the following characteristics:

Range/Distances The maximum distances, that should be covered by the network.

Number of Nodes An interval for the expected number of nodes, that will participate in the network.

Node density Although this could be derived, the density may still vary in the area, and an additional value could give a maximum and/or a minimum node density. To simplify things, I will use a qualitative measurement with the following ranges:

Very Low:	1 node per ≥ 1000 m^2
Low:	1 node per $100 - 1000$ m^2
Medium:	1 node per $25 - 100$ m^2
High:	1 node per $5 - 25$ m^2
Very High:	≥ 1 node per 5 m^2

Possibly a more reasonable value would be to use the relative density compared to the range of the transmitters in use. Since the actual transmitters used are seldom specified and the transmission range is in fact another tunable variable, I decided to ignore this for now. There is much research work going on about the ideal transmission range and power control, finding an optimal balance between covering long distances (which leads to fewer hops) and channel contention (which is increased by long transmission ranges) cf. [92].

Indoor/Outdoor Will the network be mainly indoors or outdoors (possibly both). This will have an effect about available technologies (GPS doesn't work indoors). Also propagation of wireless signals is much different.

Obstacle Density Although this is not so easy to simulate, most simulators allow settings of the general signal propagation parameters, that are appropriate for typical obstacle densities. Therefore I want to use such a parameter. Again, I use a qualitative measure. This certainly needs to be specified more precisely.

Very Low:	Outdoor plains, flat desert
Low:	Outdoors aside from settlements with moderate vegetation
Medium:	Outdoors inside of small villages and/or dense vegetation
Dense:	Outdoors in bigger towns and cities Indoors of large buildings with few walls
Very Dense:	Outdoors, but inside of a dense city center (e.g. Manhattan and comparable areas), indoors in larger buildings like shopping centers, railway stations, etc
Packed:	Inside of office buildings, flats, houses

Gateway Requirements Many applications are even more useful, if the data can be transmitted over a gateway into other networks. The two most important gateways types will be:

- Gateways into the Internet

- Gateways to a telephone network (e.g. GSM, UMTS)

Of course a GSM network could also be used to transmit data into the Internet.

Multicast Requirements Some applications could benefit a great deal from multicasting data. If this is the case, the protocol should provide multicast services.

Mobility In my opinion, a simple degree of mobility is not enough, the actual mobility can affect the performance of a routing protocol much more than one way. Still this is another parameter to classify application scenarios. I will use the following classes in several main categories:

Mobility:

Static:	nodes do not move
Very Low:	nodes move rarely and rather slow, no sudden direction changes
Low:	nodes move occasionally and slow to moderate, very few sudden direction changes
Medium:	nodes move now and then and can move slow or fast, occasional sudden direction changes
High:	nodes move most of the time, can move slow or fast, frequent direction changes
Very High:	nodes move all of the time, most of the time fast or very fast, sudden direction changes may occur very frequently

Order:

Strictly Ordered:	nodes move highly predictable/predetermined, example: trains on schedule
Mostly Ordered:	nodes move predictable and mostly predetermined examples: cars on a highway with constant conditions, trains or underground trains
Somewhat ordered:	nodes move somewhat predictable and determined, examples: cars on roads, joggers on a sidewalk, bicycles on roads
Somewhat random:	nodes move more random than predictable, examples: cars in a town center or with frequently changing traffic conditions, pedestrians in a pedestrian area
Mostly Random:	nodes move very random, examples: people on a trade fair, shoppers in a shopping center
Very Random:	Brownian to chaotic movement

Initiative:

None:	nodes cannot move by themselves, they may move due to external influences, example: sensors
Little:	nodes may move by themselves, but do it rarely, examples: sensors that occasionally correct their position, people part of a large group with high discipline
Medium:	nodes move in a group, but make independent moves as well
High:	nodes move individually most of the time, rarely in groups
Very High:	nodes never move in groups but always individual and independent

Service Requirements Some scenarios would require special services and have special requirements, some examples follow.

- Location Service

- Guaranteed Reliability

- Guaranteed Data Connections with a special set of nodes

Strategy This characteristic was defined to allow a more easy comparison with simulation scenarios in case the three (more fine grained) characteristics **Mobility**, **Order** and **Initiative** menionted above, are not availble. In my opinion the movement is better described by those mobility characteristics above than by this more simple strategy characteristic. Nevertheless it is required for the comparison, because the fine grained characteristics cannot be determined properly for most simulation scenarios.

Strategy:

Static:	nodes move barely, maybe only due to external influences or in very simple moves (e.g. straight lines)
Medium:	nodes may make more complex, but still rather simple moves, movement changes happen only occasional
High:	movement changes may happen frequently
Complex:	nodes move according to complex patterns, different nodes may move also very different, movement changes may happen frequently

Traffic Requirements Different Applications require different types of traffic. Although traffic models are complex and traffic is difficult to classify, I will provide such a classification.

I distinguish between amount and type of traffic.

Amount/Rate:

Very Low:	$0 - 2.4$ kbps
Low:	$2.4 - 19.2$ kbps
Medium:	$19.2 - 128$ kbps

High: 128 kbps -2 mbps

Very High: > 2 mbps

Type:

CBR:	audio/video/multimedia applications
interactive:	interactive sessions, chat
short file transfer:	e-mail, short messages (SMS), sensor data
medium file transfer:	web browsing
big file transfer:	web browsing, downloads
misc:	everything else

Distribution:

outgoing:	few senders, many receivers
incoming:	many senders, few receivers
balanced:	roughly equal number of senders and receivers

5.3   Matching Simulated Scenarios to Applications

In Chapter 4, I have summarized scenarios, that have been used in previous simulation based evaluations. In the following section, I will match the simulation scenarios to the application scenarios, that have been discussed previously in this chapter. This will show what kind of simulations provide best conditions in terms of applicability.

As introduced in chapter 3, in order to get a rating, I have designed a set of functions, that compare the characteristics of an application with those of a simulation scenario. Since the characteristics do not map one on one, the functions will use a rough fuzzy-like comparison in cases, where no direct comparison is possible. In some cases characteristics have greatly been simplified or even entirely omitted.

I will now briefly describe the comparisons done by the set of functions, to match application scenarios to simulation scenarios:

5.3.1 Comparison of Application and Simulation Scenarios

The requirements of the previously described application scenarios are matched against the characteristics of the simulation scenarios. This shows which simulation scenarios match best a real application and are thus most useful.

The comparison works almost the same as in the comparison between simulation scenarios in chapter 4. Some modifications are done, because the proposed characteristics of the sample applications do not directly match the characteristics of the simulation scenarios, due to the fact, that the sample application specifications are more detailed in some ways.

The reference values of the sample applications have been described earlier in this chapter in section 5.2.

Each individual comparison is also multiplied by a weightfactor and the overall result is divided by the sum of weights. The weights are chosen to emphasize on the area dimensions and node density, as well as strategy and traffic rate, as I believe these will have the biggest impact on simulation results. Obstacles and traffic type have lower weights.

Area Range The "Range/Distance" parameter of the sample application scenario is matched to the "Area" parameter of a simulation scenario. The "Range/Distance" parameter is given as an interval of possible distances $[d'_1 : d'_2]$ common for the particular application. The "Area" parameter of a simulation scenario is given as length times width $l \times w$ (without loss of generality it is assumed, that $w \leq l$). The diameter d of the area is used for the comparison and can of course be derived from length and width.

The aim is, to yield a good matching result only, if the simulation ranges are reasonably well within the specified interval.

If the width w (smaller value) is smaller than the low end of the interval d'_1, 0 is returned as result.

Else, if the diameter d of the area is at least as long as the upper end of the interval d'_2 and the larger side of the rectangle l is less than twice upper end of the interval d'_2, then a maximum match is achieved and 1.0 is returned.

If the diameter d is less than the upper end of the interval d'_2, then the following value applies:

$$\frac{d - d'_1}{d'_2 - d'_1}$$

Thus, if d is close to d'_2 we would still get a value close to 1.0.

If l is larger than two times the upper range d_2', then the excess, plus the possible excess of w is added and used as parameter in an exponential function to yield gradually smaller results.

$$\Delta l = l - 2d_2'$$

$$\Delta w = \begin{cases} w - 2d_2' & \text{if} \quad w > 2d_2' \\ 0 & \text{else} \end{cases}$$

is used in an exponential function to yield a smaller value the larger the excess:

$$e^{-\frac{10(\Delta l + \Delta w)}{4d_2'}}$$

So the result is determined as follows:

$$f_1 : \mathbb{R}_0^+ \times \mathbb{R}_0^+ \times \mathbb{R}_0^+ \times \mathbb{R}_0^+ \mapsto [0:1]$$

$$f_1(w, l, d_1', d_2') = \begin{cases} 0 & \text{if} \quad w < d_1' \\ 1.0 & \text{if} \quad w \geq d_1' \wedge d \geq d_2' \wedge l < 2d_2' \\ \frac{d - d_1'}{d_2' - d_1'} & \text{if} \quad w \geq d_1' \wedge d < d_2' \\ e^{-\frac{10(\Delta l + \Delta w)}{4d_2'}} & \text{if} \quad w \geq d_1' \wedge d \geq d_2 \wedge l \geq 2d_2' \end{cases}$$

This characteristic has a weight of 4.

Number of Nodes This is a straightforward match. If the number of nodes n_s in the simulation is within the specified interval $[n_{amin}' : n_{amax}']$ of the application, an exact match is achieved and 1.0 is returned. Else the value is $1-$ the excess value divided by the next boundary as decribed in following formula:

$$f_2 : \mathbb{N} \cup \{?\} \times \mathbb{N}^2 \mapsto [0:1]$$

$$f_2(n_s, n_{amin}', n_{amax}') = \begin{cases} 1.0 & \text{if} \quad n_{amin}' \leq n_s \leq n_{amax}' \\ 0.5 & \text{if} \quad n_s = ? \\ 1 - \frac{n_{amin}' - n_s}{n_{amin}'} & \text{if} \quad n_s < n_{amin}' \\ 1 - \frac{n_s - n_{amax}'}{n_{amax}'} & \text{if} \quad n_s > n_{amax}' \end{cases}$$

This characteristic has a weight of 3.

Node density The density characteristic from the applications will be matched against the density of the simulations, which is derived from the size of the area and the number of nodes.

The density for the application scenarios is given in terms of "very high", "medium" or "low", which correspond to intervals, as described in the beginning of this chapter (cf. section 5.2).

The density of each simulation scenario δ_{sim} (given in nodes per m^2) is compared to the density range $[\delta'_{low} : \delta'_{high}]$ specified by the sample application.

If the value is within the range 1.0 is returned, else the factor which exceeds the interval will be used to return a reduced value.

$$f_3 : \mathbb{R}_0^+ \cup \{?\} \times \mathbb{R}_0^{+2} \mapsto [0 : 1]$$

$$f_3(\delta_{sim}, \delta'_{low}, \delta'_{high}) = \begin{cases} 1.0 & \text{if } \delta'_{low} \leq \delta_{sim} \leq \delta'_{high} \\ 0.5 & \text{if } \delta_{sim} =? \\ \frac{\delta_{sim}}{\delta'_{low}} & \text{if } \delta_{sim} < \delta'_{low} \\ \frac{\delta'_{high}}{\delta_{sim}} & \text{if } \delta_{sim} > \delta'_{high} \end{cases}$$

This characteristic has a weight of 5.

Obstacles If an application requires to take obstacles into account ($\alpha_o =$ "Y"), it is checked if the simulation to be matched to, does also use movement $\alpha'_{mo} =$ "Y", or radio obstacles. $\alpha'_{ro} =$ "Y" Only if both are used, a perfect match of 1.0 is achieved, 0.5 for one type of obstacles and 0 for none.

$$f_4 : \{Y, N\} \times \{Y, N\}^2 \mapsto [0 : 1]$$

$$f_4(\alpha_o, \alpha'_{mo}, \alpha'_{ro}) = \begin{cases} 1.0 & \text{if} & \alpha_o = \alpha'_{mo} = \alpha'_{ro} \\ 0.5 & \text{if} & \alpha_o = \alpha'_{mo} \wedge \alpha_o \neq \alpha'_{ro} \vee \alpha_o = \alpha'_{ro} \wedge \alpha_o \neq \alpha'_{mo} \\ 0 & \text{else} \end{cases}$$

This characteristic has a weight of 2.

Strategy The application scenarios do not really provide a movement strategy, but their behaviour is described by the three mobility parameters (mobility, order and initiative, cf. section 5.2). I have chosen to derive an additional characteristic from these with the values: "Complex", "High", "Medium" and "Static". These should match the overall mobility and complexity of the movements typically for an application.

$$\mathbb{D}_5 = \{Complex, High, Medium, Static\}$$

$$\mathbb{D}'_5 = \{Complex, RandomWaypoint, RandomDirection, ?\}$$

$$f_5 : \mathbb{D}_5 \times \mathbb{D}'_5 \mapsto [0:1]$$

$f_5(\beta_{ms-app}, \beta_{ms-sim})$	Complex	High	Medium	Static
Complex	1.0	0.8	0.4	0.2
Random Waypoint	0.6	0.6	0.9	0.7
Random Direction	0.2	0.3	0.6	0.7
?	0.5	0.5	0.5	0.5

This characteristic has a weight of 4.

Traffic rate The traffic rate of the application scenarios is only classified as "Low", "Medium" or "High". The traffic rate r in kbit/s given in the simulation scenario, (computed from the individual rate and the number of sources) is then matched against this classification using the following matrix:

$$f_6 : \{Low, Medium, High\} \times \mathbb{R}_0^+ \cup \{?\} \mapsto [0:1]$$

$f_6(r_{tapp}, r_{tsim})$	Low	Medium	High
$r =$ "?"	0.5	0.5	0.5
$r < 2.4$	0.8	0.3	0.05
$2.4 \leq r < 19.2$	1.0	0.8	0.2
$19.2 \leq r < 128$	0.2	1.0	0.5
$128 \leq r < 2000$	0.05	0.5	1.0
$r \geq 2000$	0	0.05	0.5

This characteristic has a weight of 4.

Traffic type The type of traffic of the application is matched against the type used in evaluations. This match is not really interesting, since all considered simulations just use CBR/UDP traffic, while applications make use of different types of traffic. As described in section 5.2, ithe applications scenario use the following classification of traffic requirements: "CBR", "Interactive", "File Transfer" and "misc" (actually there are different variants of file transfer, but this has been simplified for the comparison).

Again a fixed assignment of values is used:

$$\mathbb{D}_7 = \{CBR, Interactive, FileTransfer, Misc\}$$

$$\mathbb{D}'_7 = \{CBR/UDP, VBR/TCP, ?\}$$

$$f_7 : \mathbb{D}_7 \times \mathbb{D}'_7 \mapsto [0 : 1]$$

$f_7(\beta_{ttapp}, \beta_{ttsim})$	CBR	Interactive	File Transfer	Misc
CBR/UDP	1.0	0	0.2	0.5
VBR/TCP	0	0.8	0.8	0.5
?	0.5	0.5	0.5	0.5

This characteristic has a weight of 2.

Thus the overall matching result can be computed as follows:

$$f = \frac{1}{24}(4f_1 + 3f_2 + 5f_3 + 2f_4 + 4f_5 + 4f_6 + 2f_7)$$

5.3.2 Example

The following example will illustrate how the matching works. The Johanssen "Conference Room" scenario [85] will be used as simulation scenario and it will be compared to the "Conference Room" (cf. section 5.5.2) application scenario in table 5.1 and also to the "Event Coverage" (cf. section 5.5.4) application scenario in table 5.2.

Characteristic	Value Sim.	Value App.	Weight	Result
Area-Range	$w = 90$ $l = 150$ $d \approx 175$	$d'_1 = 10$ $d'_2 = 100$	4	4.0
Number of Nodes	$n_s = 50$	$n_{amin} = 50$ $n_{amax} = 300$	3	3.0
Node Density	$\delta_{sim} = \frac{1}{270}$	$\delta'_{low} = \frac{1}{5}$ $\delta'_{high} = \infty$	5	0.09
Obstacles	$\alpha'_{mo} = Y, \alpha'_{ro} = Y$	$\alpha_o = Y$	2	2.0
Strategy	$\beta_{ms-sim} = $ Complex	$\beta_{ms-app} = $ Static	4	0.8
Traffic Rate	$r_{tsim} = 120$	$r_{tapp} = $ Low	4	0.8
Traffic Type	$\beta_{ttsim} = $ CBR/UDP	$\beta_{ttapp} = $ CBR	2	2.0
Sum:			24	12.69
Result:				0.53

Table 5.1: Conference Room Simulation - Conference Room Application

Characteristic	Value Sim.	Value App.	Weight	Result
Area-Range	$w = 90$ $l = 150$ $d \approx 175$	$d'_1 = 200$ $d'_2 = 1000$	4	0
Number of Nodes	$n_s = 50$	$n_{amin} = 500$ $n_{amax} = 20000$	3	0.3
Node Density	$\delta_{sim} = \frac{1}{270}$	$\delta'_{low} = \frac{1}{5}$ $\delta'_{high} = \infty$	5	0.09
Obstacles	$\alpha'_{mo} = Y, \alpha'_{ro} = Y$	$\alpha_o = N$	2	0
Strategy	$\beta_{ms-sim} = $ Complex	$\beta_{ms-app} = $ Static	4	0.8
Traffic Rate	$r_{tsim} = 120$	$r_{tapp} = $ Medium	4	4.0
Traffic Type	$\beta_{ttsim} = $ CBR/UDP	$\beta_{ttapp} = $ CBR	2	2.0
Sum:			24	7.19
Result:				0.30

Table 5.2: Conference Room Simulation - Event Coverage Application

5.4   Matching Results Overview

The following graph 5.1 shows the average matching of an simulation scenario to all application types. This means, that the matching values for each characteristic of all application scenarios are averaged for a single simulation scenario. Thus an average matching value for each simulation scenario is obtained. This would show, if any particular simulation scenario is specially suited for a broad range of applications. As the graph shows, there is no particular simulation scenario which would be suitable for a larger set of applications (none $\geq 60\%$).

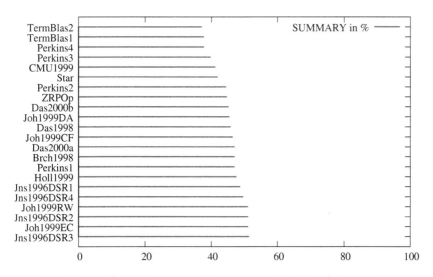

Figure 5.1: Average Matching of Evaluations to all Applications

The following figure 5.2 shows a more direct mapping from applications (in the top row) to the simulations (in the bottom row). Only matchings $\geq 70\%$ are displayed in order to keep the diagram clearly arranged.

The "Park Management" application is matched well by the Johansson Random Waypoint scenario as well as the Star simulation experiment mainly because of commonly low node density. Office building and personal area network applications are matched very well by the Johnson simulations. This is due to the fact, that these experiments were restricted to a very small area and thus also have a high node density, which are typical characteristics for these two applications. Interestingly the car based networks application is also matched well by Johnson simulations. With car based networks, I have also assumed a high node density (as it will be typical for cars in large cities or on heavily used highways).

The following two tables show all the matching results with the better values highlighted[1].

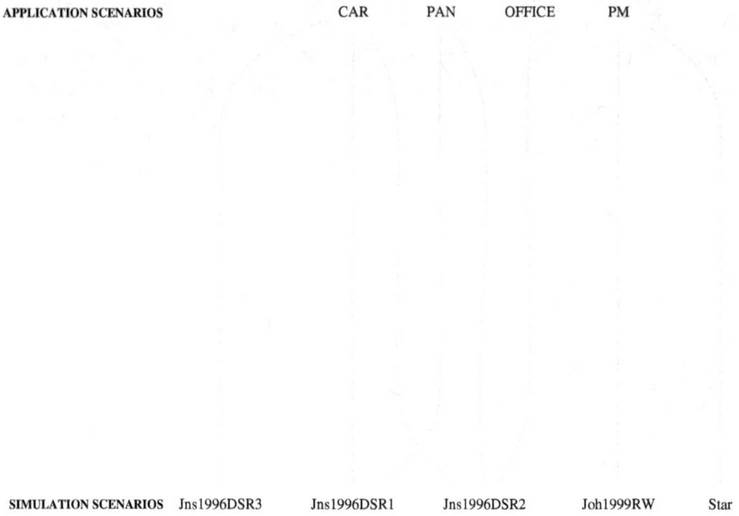

APPLICATION SCENARIOS	CAR	PAN	OFFICE	PM

SIMULATION SCENARIOS	Jns1996DSR3	Jns1996DSR1	Jns1996DSR2	Joh1999RW	Star

Figure 5.2: Matching Applications to Evaluations $\geq 70\%$

[1]Please note that due to rounding, some border values (e.g. 70%) may appear in different colors in the tables.

Sims/Apps	CAR	CELL	CR	DA	EC	ISN
Brch1998	0.56	0.35	0.33	0.32	0.55	0.56
CMU1999	0.38	0.45	0.26	0.43	0.33	0.42
Das1998	0.49	0.50	0.33	0.25	0.55	0.49
Das2000a	0.56	0.35	0.33	0.32	0.55	0.56
Das2000b	0.56	0.34	0.33	0.37	0.49	0.61
Holl1999	0.48	0.35	0.33	0.47	0.65	0.53
Jns1996DSR1		0.36	0.30	0.25	0.45	0.60
Jns1996DSR2		0.34	0.39	0.21	0.52	0.65
Jns1996DSR3		0.36	0.46	0.21	0.58	0.63
Jns1996DSR4	0.70	0.36	0.48	0.21	0.58	0.58
Joh1999CF	0.48	0.49	0.53	0.50	0.30	0.54
Joh1999DA	0.54	0.35	0.33	0.48	0.38	0.58
Joh1999EC	0.39	0.51	0.49	0.49	0.46	0.66
Joh1999RW	0.48	0.53	0.36	0.34	0.63	0.63
Perkins1	0.41	0.58	0.41	0.32	0.55	0.49
Perkins2	0.41	0.41	0.41	0.37	0.56	0.56
Perkins3	0.41	0.41	0.41	0.43	0.44	0.56
Perkins4	0.42	0.28	0.33	0.43	0.49	0.56
Star	0.49	0.41	0.33	0.32	0.37	0.44
TermBlas1	0.46	0.20	0.28	0.43	0.52	0.63
TermBlas2	0.47	0.20	0.28	0.43	0.52	0.55
ZRPOp	0.41	0.58	0.41	0.32	0.58	0.49

Sims/Apps	MIL	OFFICE	PAN	PM	SN	TF
Brch1998	0.34	0.49	0.33	0.68	0.55	0.58
CMU1999	0.59	0.34	0.37	0.50	0.45	0.44
Das1998	0.40	0.49	0.33	0.63	0.50	0.53
Das2000a	0.34	0.49	0.33	0.68	0.55	0.58
Das2000b	0.20	0.49	0.33	0.68	0.48	0.53
Holl1999	0.43	0.42	0.33	0.50	0.65	0.57
Jns1996DSR1	0.35	0.65	0.73	0.41	0.54	0.51
Jns1996DSR2	0.33	0.70	0.71	0.46	0.59	0.46
Jns1996DSR3	0.35	0.68	0.69	0.46	0.57	0.44
Jns1996DSR4	0.35	0.63	0.64	0.46	0.52	0.43
Joh1999CF	0.69	0.52	0.38	0.43	0.29	0.42
Joh1999DA	0.49	0.49	0.33	0.51	0.38	0.58
Joh1999EC	0.42	0.49	0.49	0.60	0.54	0.58
Joh1999RW	0.58	0.41	0.36		0.58	0.50
Perkins1	0.43	0.41	0.41	0.63	0.58	0.45
Perkins2	0.14	0.41	0.41	0.63	0.58	0.46
Perkins3	0.13	0.28	0.28	0.64	0.43	0.34
Perkins4	0.13	0.28	0.28	0.53	0.41	0.39
Star	0.33	0.41	0.41		0.46	0.32
TermBlas1	0.27	0.25	0.20	0.59	0.34	0.35
TermBlas2	0.27	0.25	0.20	0.61	0.34	0.35
ZRPOp	0.30	0.28	0.28	0.63	0.58	0.49

5.5 Application Scenarios in Detail

In this section all application scenarios will be described with their characteristics. Further a diagram will show how well the simulation scenarios described in the previous chapter 4 will match each application according to my comparison functions. The details of the matching function, implemented in a perl script are explained in the previous section 5.3.

5.5.1 PANs and Bluetooth

A personal area network is very short of range and connects devices attached to personal clothing, PDAs, etc. But with such small distances, it seems not a problem to connect each device directly. Each device should be in range of another even with short range transmitters. If a stationary node should be contacted by the PAN, e.g. to establish an Internet connection, again, this would not require a MANET.

This could change if inter-person communication should also be possible. In this case this type application will match more closely a low scale version of the *Individual Spontaneous Networks* in section 5.5.7.

Bluetooth was designed to eliminate wires from home computing and possibly home entertainment devices and could even connect home appliances. A Bluetooth *piconet* is a well organized but limited network. However, piconets can be interconnected into a *scatternet*. Scatternets work in a self organized ad hoc network fashion and could be used to connect more devices in a house or flat and even devices in a PAN.

Scatternets are part of the research of the MANET community, but are not part of my research. Also the choice of protocol is already determined.

Characteristics of a PAN application would be:

Range: ≤ 10m
No of Nodes: ≪ 100
Node Density: High
Sky: Indoor
Obstacle Density: Packed
Gateway Requirements: none particular, could be used
Multicast: not required
Mobility: Low Mobility
Order: Mostly Random
Initiative: Little
Strategy: Static
Traffic amount: Low
Traffic type: CBR?
Traffic distribution: balanced

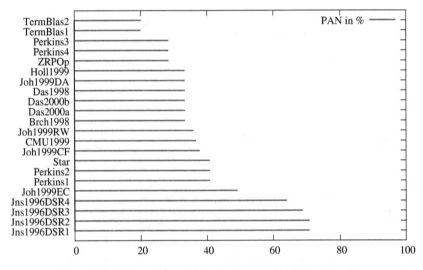

Figure 5.3: Matching of Simulations to PAN Application

Figure 5.3 shows that the early DSR (cf. section 4.7.2) experiments match best. This is expected, since these experiments are characterized by very short range transmissions and a small room.

5.5.2 Conference Room

The scenario of a conference room is a more advanced of the proposed scenarios. A speaker offers some data and the audience may consume that data and may also exchange information with other members of the audience. Movement patterns are quite distinct.

In my opinion, the conference room is not a typical application for mobile ad hoc networks. It is more a classical WaveLAN application. Conferences usually happen at dedicated places, so it is likely that WaveLAN infrastructure either exists, or can easily be established. Further even small PDAs are nowadays capable of WaveLAN functionality. It is a well established standard and offers reasonable performance for data to be exchanged in such an occasion.

I guess there are few reasons to choose a multi-hop ad hoc network over a simple shared WLAN in such a situation. If it would still be decided to use ad hoc networking, the frame requirements and constraints would be:

Range:	$\leq$ 100m
No of Nodes:	50 − 300
Node Density:	Very High
Sky:	Indoor
Obstacle Density:	Dense
Gateway Requirements:	Internet desirable
Multicast:	useful
Mobility:	Very Low Mobility
Order:	Mostly Random
Initiative:	Little
Strategy:	Static
Traffic amount:	Low
Traffic type:	CBR?
Traffic distribution:	outgoing

Figure 5.4 shows that Johansson's "Conference Room" scenario (cf. section 4.7.2) matches best to my definition of a conference room. Apart from "Event Coverage" all other experiments do not match very well.

5.5.3   Trade Fair Networks

This application has many things in common with the Conference Room scenario. As well, trade fairs happen at special centers and an existing infrastructure is likely to be present. Although the distances will be larger than in a conference room and infrastructure cannot be added so easy, it appears to be likely, that convention centers will establish such infrastructure to offer network services to their exhibitors and visitors. Thus the need for ad hoc networking in such situations is greatly reduced.

Range:	≤ 1000m
No of Nodes:	500 – 3000
Node Density:	Medium
Sky:	Indoor(Outdoor)
Obstacle Density:	Dense
Gateway Requirements:	Internet desirable
Multicast:	very useful
Mobility:	Medium
Order:	Somewhat Random
Initiative:	Medium
Strategy:	High
Traffic amount:	High
Traffic type:	CBR, File Transfer
Traffic distribution:	outgoing

Figure 5.5 shows, that even four simulation scenarios match equally well. These are the Johanssen Disaster Area and Event Coverage scenario, which is expected, because the Trade Fair application scenario was modeled after the Event Coverage scenario, also these two simulation scenarios are rather similar. Further experiments of Broch et al 1998 and the first simulation of Das, Perkins and Royer 2000 match equally well. Those experiments have

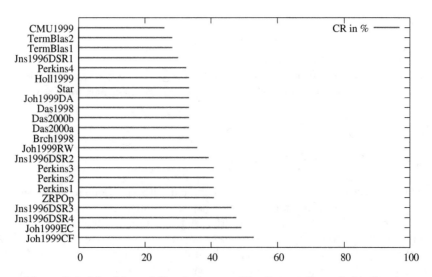

Figure 5.4: Matching of Simulations to "Conference Room" Application

again a high degree of similarity to each other, as shown in diagram 4.17 in section 4.7.3.

5.5.4 Event Coverage

An "Event Coverage" scenario was first proposed in [85]. However, it appeared that this was not as general as the description would imply. Thus I split up the types of events into trade fair and exhibition type events (cf. section 5.5.3) and events like music festivals/rock concerts, open-air cinemas and shareholder meetings. The latter could be imagined as examples for my type of "Event Coverage" application. Although the movement and node distribution is related to the Trade Fair application, the communication patterns will be different. Also it is likely (especially for open-air events) that it would require much more effort to establish a WaveLAN infrastructure. As noted, my "Event Coverage" application scenario differs significantly from the "Event Coverage" evaluation scenario as described in the evaluations by Per Johansson [85], cf. section 4.2.3. The previous "Trade Fair" application scenario (section 5.5.3) is much closer to Johansson's definition.

A typical scenario would have the following characteristics.

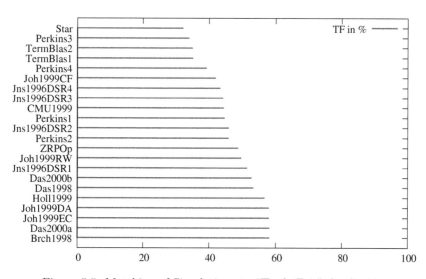

Figure 5.5: Matching of Simulations to "Trade Fair" Application

Range:	200 − 1000m
No of Nodes:	500 − 20000
Node Density:	Very High
Sky:	Outdoor(Indoor)
Obstacle Density:	Low
Gateway Requirements:	none special
Multicast:	useful
Mobility:	Low
Order:	Mostly Random
Initiative:	Medium
Strategy:	Medium
Traffic amount:	Medium
Traffic type:	CBR, Interactive
Traffic distribution:	outgoing-balanced

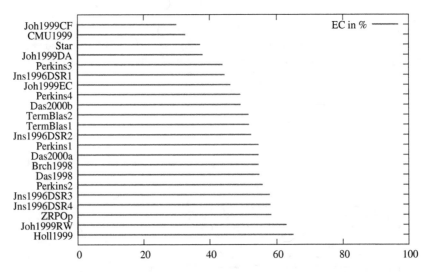

Figure 5.6: Matching of Simulations to "Event Coverage" Application

The differences to Johansson's "Event Coverage" are also illustrated in figure
5.6, since his experiment does not match very well. Indeed with less than
50% it is one of the least matching experiments. Instead the experiment by
Holland and Vaidya (cf. section 4.7.2) matches best. This is only the case,
because this experiment is described most unspecific, so many parameters
are unknown. The way the matching function works, unknown parameters
always match to some degree, resulting in an apparently good match, but
which has to be considered with care. Johansson's modified Random Way-

point experiment matches nearly as good, in this case because the movement patterns and obstacle requirements match.

5.5.5 Extended Cellular Phone Networks

One example application which is mentioned frequently is to extend a cellular phone network. If either the base station is too far away, or the cell has reached capacity limits, still people could connect by using another user as a relay.

I believe that cell coverage is very much complete in cellular networks these days, so the main issue would be to remedy cell contention. In this case bandwidth for the node with the connection would be reduced. I have doubts, if ad hoc networking could solve such problems.

If one wants to explore the situation of extended cell coverage, there would be a special situation:

Range:	$10 - 500$m
No of Nodes:	$20 - 200$
Node Density:	Medium
Sky:	Outdoor(Indoor)
Obstacle Density:	Medium
Gateway Requirements:	GSM or UMTS required
Multicast:	not useful
Mobility:	Medium
Order:	Mostly Random
Initiative:	Medium
Strategy:	Static
Traffic amount:	Low
Traffic type:	CBR
Traffic distribution:	outgoing and incoming

Special Considerations

- All communication has a fixed node (BTS) as source or sink. So the data does not go to any node, but to a single node once the connection is established.

- Each node wants to connect either directly or indirectly to the closest BTS node.

- A node with no connection wants to use a node as a relay that has a connection with few hops as possible.

- GSM/UMTS services need to be supported.

- Accounting may be required.

Since this is a very special type of application and it's special considerations cannot be taken into account while doing the comparison and in fact none of the previous experiments actually tried to model such an application. There is no simulation experiment that matches better than 60% (cf. figure 5.7), but the ZRP simulations (Haas and Pearlman 1999) and Perkins' and Royer's (2001) first experiment match best. Again these experiments are similar to a high degree (78%).

5.5.6 Office Building Networks

Office Buildings nowadays most likely have a wired infrastructure present. Thus it would be very easy to add a WaveLAN infrastructure if required.

Further ad hoc networking would have to cope with many obstacles obstructing communications, as there are many walls, ceilings and floors made of concrete which usually inhibit communications beyond the same room.

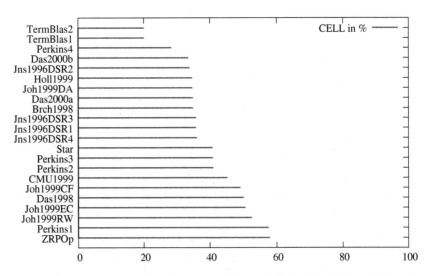

Figure 5.7: Matching of Simulations to "Extended Cellular Network" Application

Such an environment would be very difficult to use for simulation. The benefits of an ad hoc network in such an environment are doubtful, since a static or WLAN network infrastructure could be deployed very easily and ist probably already present in many places. Thus, I don't consider office buildings as a key sample application for mobile ad hoc networks.

Range:	300m
No of Nodes:	100
Node Density:	High
Sky:	Indoor
Obstacle Density:	Packed
Gateway Requirements:	Internet / intranet required
Multicast:	not very useful
Mobility:	Very Low
Order:	Somewhat Random
Initiative:	Medium
Strategy:	Static
Traffic amount:	High
Traffic type:	mixed
Traffic distribution:	balanced

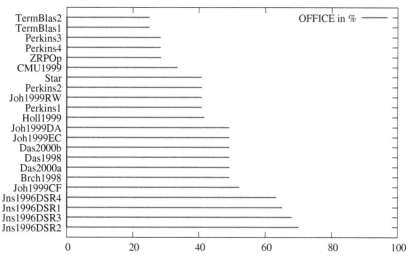

Figure 5.8: Matching of Simulations to "Office Building Network" Application

Due to the high density and short ranges, the Johnson experiments (cf. section 4.7.2) match best, while the wide area and long range experiments of the Terminode project (cf. section 4.7.2) match least (cf. figure 5.8).

5.5.7   Individual Spontaneous Networks

This should describe the *vision* that some promoters of mobile ad hoc networks are advertising: People in a large area are able to communicate without relying on an existing infrastructure of a large carrier and also without the need to pay any fees. Anyone could be part of such a large scale network and use it. If a node does not have own communication in progress, it can be used as a relay for other nodes.

Such networks cover a large scale. From a few pedestrians in a neighborhood up to a crowded city center or even beyond town boundaries to cover a whole region (like suggested by the Terminode project [154]). Cars, buses, cyclists, trains, streetcar, etc. can take part in the network.

This is one of the most complex applications for mobile ad hoc networks. The characteristics and requirements are difficult to pinpoint here.

Range:	$500 - 20000$m
No of Nodes:	$10 - 10000$
Node Density:	High
Sky:	Outdoor
Obstacle Density:	High
Gateway Requirements:	Internet desirable
Multicast:	not very useful
Mobility:	Medium
Order:	Mostly Random
Initiative:	High
Strategy:	Medium
Traffic amount:	Medium
Traffic type:	mixed
Traffic distribution:	balanced

The number of nodes can range from 10 to 10.000 maybe even more. The area can vary from a few meters across, up to several kilometers. Too short ranges probably don't make too much sense (since a single hop WLAN would be sufficient then), so an example setting would be 500-20000 meters.

There are several different types of nodes with different behavior and characteristics. Traffic is difficult to predict. Also static Internet gateway nodes are likely part of such a network.

Special Considerations:

- The above parameters are subject to a high variation. Thus the underlying routing framework would need to be highly scalable and adaptive.

- The application data transmitted will also be of a large variety: video, voice, chat, email, file transfer, web browsing, etc.

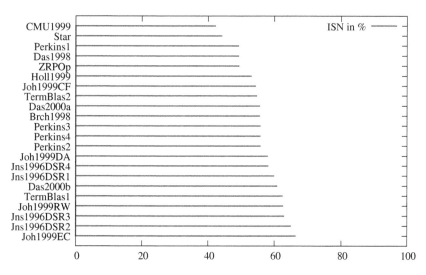

Figure 5.9: Matching of Simulations to "Individual Spontaneous Network" Application

There is no clear distinct best match for this type of application, as seen in figure 5.9. Johansson's "Event Coverage" experiment (cf. section 4.7.2) as well as teo of the Johnson experiments are in the front of the best matching experiments. The intervals of this application scenario are kept very wide and thus the requirements are very general. Thus a lot of experiments achieve to match better than 50%. They high node density will allow the Johnson experiments to be in the top 3, while the use of obstacles is an advantage for the Johansson event coverage experiment.

5.5.8 Car Based Networks

Car based networks are proposed to distribute information regarding cars and traffic. Congestion warnings as well as other special situations (accident, changing weather conditions like ice, fog, etc) could spread quickly between cars, much better than the usual radio station could distribute such information.

Further it is possible to relay information from static nodes which could be placed along the roads and highways. Thus, independent companies can offer additional information for the travelers.

Cars have even more advantages for mobile ad hoc networks, since they have a virtual unlimited power source and could thus even provide much higher transmission power.

Range:	$1000 - 100000$m
No of Nodes:	$10 - 1000$
Node Density:	High
Sky:	Outdoor
Obstacle Density:	Very Low
Gateway Requirements:	Internet desirable
Multicast:	useful
Mobility:	High
Order:	Mostly Ordered
Initiative:	Medium
Strategy:	High
Traffic amount:	High
Traffic type:	CBR,file transfer
Traffic distribution:	outgoing

Special Considerations:

- Movement patterns on a road are distinct and there exist models for such movement behavior[71]. These could tend to long routes.

- Communication is centered about certain hot spots, e.g. cars close to an accident or a traffic jam, roadside fixed nodes with special services, etc.

- Services provided could include: video, voice, chat, traffic-information, news, etc. So communications takes place between cars and special fixed nodes, but also between cars (as originator for data).

This time the figure 5.10 shows a paradox result. The short range experiments by Johnson, which have nothing to do with cars, seem to match best, while just the one single testbed experiment, that used real cars (cf. 4.7.2) did match worst. The explanation is, that node density, number of nodes and traffic rate match exceptionally well in this combination. This even compensates the fact, that the range does not match at all. The real car testbed instead had a very low node density, and an even too low number of nodes.

This shows, that even apparently similar experiments and application can differ a great deal.

5.5.9 Farm and Park Management

Any project that works under an open sky and in the nature could benefit from mobile ad hoc networks. In a crop field it is undesirable to install base stations or any other wireless communication infrastructure. The same applies for nature parks or national parks. Still the work in such areas could benefit from data communication. Sensor or sample data could quickly be transmitted to coworkers or relayed to a base or park or farm headquarters.

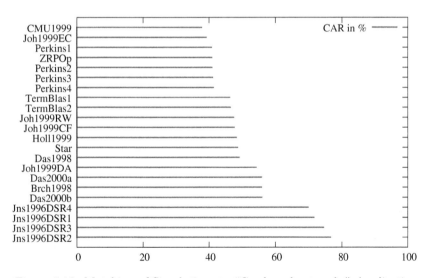

Figure 5.10: Matching of Simulations to "Car based networks" Application

Range: $100 - 20000$m
No of Nodes: $10 - 200$
Node Density: Very Low
Sky: Outdoor
Obstacle Density: Very Low
Gateway Requirements: none
Multicast: not very useful
Mobility: Low
Order: Mostly Ordered
Initiative: Medium
Strategy: Medium
Traffic amount: Medium
Traffic type: file transfer
Traffic distribution: incoming

Special Considerations:

- There will be a few fixed nodes, which represent e.g. farm headquarters or ranger stations.

- Services will include: data, chat, email/short messages and possibly voice.

- However, in practice my experience is very limited, thus the data requirements are very vague.

Again the node density weight favors the best matching experiments in figure 5.11. The low density is clearly the most prominent parameter of this application, thus Johanssons Random Waypoint and the Star experiment yield the best matches.

5.5.10   Sensor Networks

This is a special application for ad hoc networks. Sensors are typically used in areas, where no infrastructure is present or cannot be established quickly but where data needs to be collected.
Sensors can be much more efficient, if they can coordinate themselves. Further, the collected data must be transmitted to a collecting device (a distinct node).
Sensors usually have a limited power source and very little autonomous movement abilities. Sensor networks are an important part of MANET research and special set of communication protocols. I did not take sensor networks into account of this study in order to limit the amount of considered methods.

Characteristics include, that nodes do not move on their own, but can drift. The communication range is limited. A small set of distinct nodes will collect the data and therefore act as sink.

Range:	$10 - 1000$m
No of Nodes:	$10 - 1000$
Node Density:	High
Sky:	Outdoor
Obstacle Density:	Low
Gateway Requirements:	none
Multicast:	not very useful
Mobility:	Low
Order:	Mostly Ordered
Initiative:	None
Strategy:	Medium
Traffic amount:	Low
Traffic type:	file transfer
Traffic distribution:	incoming

Special Considerations:

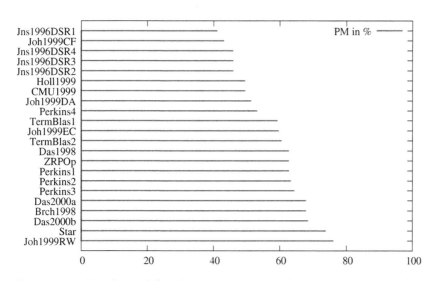

Figure 5.11: Matching of Simulations to "Farm and Park Management" Application

- Nodes rather drift than move on their own.

- Communication range will be limited due to power restrictions.

- There will be a small set of nodes, which act as receivers (data collectors) for the majority.

- Services: sensor data

- Power consumption is an issue.

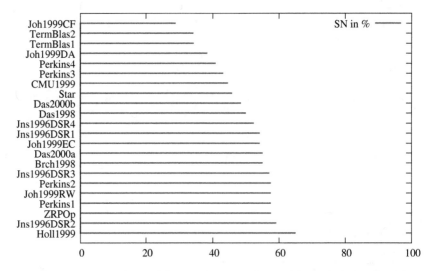

Figure 5.12: Matching of Simulations to "Sensor Networks" Application

My choice of routing protocols and simulation experiments to be considered did not include the special area of sensor networks [2], consequently none of the experiments matches very well to the sensor networks application, as shown in figure 5.12. Only the Holland and Vaidya experiment has a better match, but again, only because it has so many unknown values.

5.5.11   Disaster Area Recovery Support

Another typical situation is a site of an catastrophic event. There are many people to coordinate (rescue teams, paramedics, firemen, etc) and existing

---

[2]in order to limit the material to examine to an amount that could be coped with

infrastructure may be destroyed. In such a situation mobile ad hoc networks could play an important part of the necessary coordination.

This is also a rather complex scenario. I think that an ad hoc network is only worth the trouble in a big catastrophic event.

Range:	500 − 3000m
No of Nodes:	50 − 500
Node Density:	Medium
Sky:	Outdoor
Obstacle Density:	Medium
Gateway Requirements:	none
Multicast:	not very useful
Mobility:	High
Order:	Somewhat Random
Initiative:	High
Strategy:	Complex
Traffic amount:	Low
Traffic type:	file transfer
Traffic distribution:	incoming

Special Considerations:

- Different types of nodes: paramedics, firemen, fire trucks, ambulances, helicopters.

- Communication and also movement takes place around certain hot spots in the area.

- There are dangerous areas, safe areas, areas with injured people, areas with rescue infrastructure (heliports), which are all differently approached.

- Robustness and reliability is a key issue in this scenario.

- Services used will be: data, chat, voice, e-mail/messaging.

- Communication will also take place with special nodes coordinating the rescue teams (headquarters).

Interestingly, not Johansson's "Disaster Area" (cf. section 4.7.2) matches best, but Johansson's "Conference Room" scenario, but Johansson's "Disaster Area" comes close behind. In any case all these scenarios are very similar.

5.5.12   Military Applications

Military is one of the key applications for mobile ad hoc networks and the first research efforts about MANETs have been initiated by the military (starting with the PRnet).

The reasons are obvious: in the battlefield there is no infrastructure that can be used, task forces are certainly more efficient if they are coordinated and autonomous weapon systems (drones) may act as nodes in a MANET. Also infrastructure-less communication is far more robust against enemy counter-measures.

However, since I do not have any military background and or experience, I cannot speculate about the details of such scenarios.

For *Soldiers in the Field*, I would assume the following constraints:

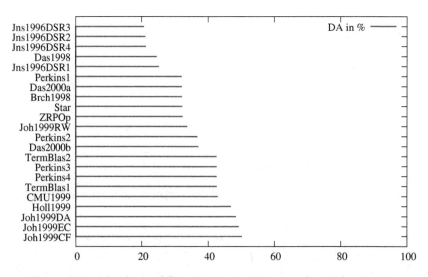

Figure 5.13: Matching of Simulations to "Disaster Area" Application

Range:	$10 - 500$m
No of Nodes:	$20 - 50$
Node Density:	Medium
Sky:	Outdoor
Obstacle Density:	Medium
Gateway Requirements:	to command
Multicast:	useful
Mobility:	Medium
Order:	Somewhat ordered
Initiative:	Medium
Strategy:	Complex
Traffic amount:	Medium
Traffic type:	interactive, file transfer
Traffic distribution:	incoming

Special Considerations:

- Not that many different types of nodes: soldiers, few officers and some kind of link to the next rank unit.

- Robustness and reliability is a key issue in this scenario.

- Transmission need to be secure and especially robust.

- Services used will be file transfer (tactical data) and possibly interactive traffic (notification of events).

Other military units may also benefit from mobile ad hoc networking, ranging from helicopters and aircraft up to naval forces. Due to lack of knowledge in this area, I do not consider them much further.

Again there is no good match except Johansson's "Conference Room", (cf. figure 5.14), the ranges and node density, as well as the movement strategy lead to this good match.

5.6 Summary

Except for the special situation, where the Johnson experiments match the CAR application very well, clearly the Johansson experiments (cf. section 4.7.2, [85]), yield good matches to many of the proposed applications. I account this to the outstanding degree of realism, which was taken care of with much effort in these experiments. There are few matches above 60% and even less above 70%. However most of the matches above 70% are sensible and justified. So this means, that the matching function may not be altogether useless.

However, the importance of such an comparison depends on the fact if the results of more realistic scenarios are significantly different, than those of much simpler scenarios under comparable conditions (e.g. same number of nodes, size of area, same type of traffic). It would be sufficient to show this for the case of the experiments in [85]. Even though in [85], a simple scenario was examined as well, the unaffected constraints have been varied without need, thus reducing the value of this study.

It is my belief, that certain aspects of realism will have a significant impact on the results of evaluations, but it is required to further investigate which aspects will affect the results most. In order to do this, simulations must be performed to compare simple and realistic scenarios with a selected set of routing protocols.

In chapter 11 some sample simulations have been performed and examined also in this respect. The results show significant differences between simple movement scenarios and more complex ones. If this can be confirmed in more thorough simulations, it is evident, that more realistic simulations scenarios, which are modeled after the intended applications have to be used in future simulations.

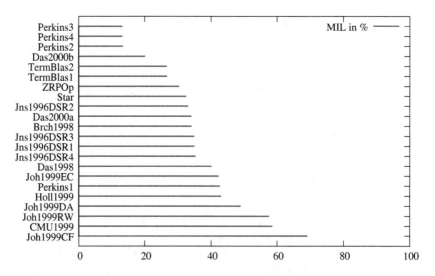

Figure 5.14: Matching of Simulations to "Military" Application

Chapter 6

Other Scenario Considerations

In this chapter, I will describe other considerations and thought about evaluation scenarios. These considerations take other aspects of realism and applicability of evaluation scenarios into account.

6.1 Other Views of Scenarios

As described, current scenarios observe a particular region of a given geometry and size, populated with a specific amount of nodes over a given time period. I will call this view an *area based model*.

If one observes a fixed area, with certain characteristics that resemble a particular location in the real world, like an office building, a popular town square or a battlefield, the movement pattern of nodes is not static, but varies over time. A typical time period that shows regular changes could be 24 hours.

Many people arrive at the office in the morning, then do their work, more or less distributed over the area, go for lunch around noon, leave the building in the evening with just a few people remaining there at night.

This example shows, that the environment in which the routing has to take place may change a great deal over time. Parameters like node density can reach peak values during a certain period of a day, so this needs to be taken into account for a decision what routing protocol may be best suited to use for such an application.

Depending on the time frame, such changes need to be considered in an *area based model* (i.e. it remains centered around a particular are with nodes coming and going, depending on the time period).

Depending on the problem, other possible views may be better suited. An

113

obvious alternative would be a *node based model*: On observing nodes, some nodes will not be present in an particular area over the observed period. Nodes may move out of this area and into other ones.

It is arguable, that the *area based model* is not correct, since the office building does not run any routing protocol at all, but the nodes do it. Thus it may be considered to choose a scenario, that is not fixed on a certain location, but on a certain node, and describes the node's environment (in terms of other nodes, density, obstacles, etc) over a certain period of time. Maybe a circle of a given radius around a fixed node, with his environment reflecting the various situations it is confronted with, could be an appropriate way.

On a first glance, such a model would be much more difficult to implement (mainly because fixed obstacles would not appear fixed but change their location, from a single node's point of view).

It makes sense to think of a way to use the easier and well understood properties of an *area based model*, together with the more realistic changing environment of the *node based model*. A possible solution would be the following suggestion, which I call *phase based model*:

It appears possible to break down the changing environment of the *node based model* into a set of situations, like *driving through town to work, entering office building, work at workplace, attend meetings, go to lunch, have lunch, leave office building, drive home from work, spend remaining time at home.*

Such a sequence of more static scenarios which could be bound to specific areas, may reflect the changing environment (as in the *node based model*) well enough. The phase transition itself is not of great scientific interest, since it is not expected to cause any unusual behavior. Consequently the *phase based model* does not reduce the scientific gain from the *node based model*, while greatly simplifying a possible implementation.

6.2   Empirical Movement Data for Scenarios

Any artificial mobility model may have problems to model real movement close enough. There must be ways to verify the applicability of a model or even better would be data to derive the model from.

Such empirical data would be a very good thing to start with. Alas, to collect a representative set of such data from possibly thousands of people at various places over a long period of time seems impossible for an individual. Mobile phone carriers will most likely have such data available, since it is possible to track mobile phone users with the known locations of the base stations and such data will be used to optimize the cell layout and placement of base stations.

6.2.1 Tracks generated by a GPS receiver

In an experiment I have tried to track my location over a long time with a GPS receiver.

Although this resulted in a variety of movement patterns, this method had a lot of drawbacks:

- GPS reception only works under a clear sky. Any obstruction by dense vegetation, tunnels or buildings results in missing and/or inaccurate data. Traveling by train, tram or bus results in bad tracking data, as well. Tracking inside of buildings is impossible.

- GPS receivers are clumsy and consume a lot of power (compared to mobile phones).

- GPS receivers are very expensive.

Nevertheless, I have collected the data and tried to analyse it, the results are presented in appendix B. It is very hard to draw conclusions from this data or to try to build a model upon them, because I have used very different means of traveling. The collected data results from traveling on foot, by car, by bicycle, by bus and by tram line. It would have been even more inconvenient to record each means of travel separately.

Regardless, I figured that the easiest way to model the recorded movements would be to implement "Smooth is better than sharp" [15]. Thus I have implemented this model and have created some statistical data to compare with the GPS data. The parameters and results are presented in appendix C. Although the graphs are not really comparable it appears that is indeed possible to use the "Smooth is better than sharp" model. The data is not really comparable, since there are too many uncertainties in the GPS data, which have not been considered, like the time intervals between data samples. (The time intervals between the samples of my implementation of the "Smooth is better than sharp" model are likewise not fixed. Each sample is based on a *move*. A *move* is continues until the Poisson process generates another one.)

Due to the impractical handling of GPS devices, their high costs and high energy consumption, data can not be collected from a representative set of persons. It was no little effort to collect the data from just myself, but a representative set of people may exceed 1000 persons. Even, if a much larger number of people could be tracked, one could still not get data like node density[1], without additional measures, like counting people passing-by, etc. Thus, I have decided not to continue that way of collection empirical data for movement patterns.

6.2.2 Data from Telecommunication Carriers

The next logical step is to look around, who else could have such data, since I can not collect it by myself. It is likely that telecommunication carriers for mobile phone networks did collect such data on a large scale, since such tracking data of mobile phone users is crucial for the layout and structure of communication cells. Also it is easy to track mobile phone users, since the transceiver base stations can act as reference points with a known location[2]. Unfortunately, phone companies do not seem to give away such data to the public and not even to the research community. My attempts to get hold of such tracking data have been rejected. Thus this helpful start is unfortunately not available.

The conclusion is, if someone is able to get representative empirical data about node distribution and movement, it would be a valuable input for designing application aware scenarios for evaluation of mobile ad hoc networks.

6.3 Movement Properties

How are nodes supposed to move then if not as described in the scenarios already used. I assume that each node is attached to a person or a vehicle, that is controlled by a person[3].

If one looks at an individual node, it's movement can be described by two key properties.

Kinetical State I use this term to describe the triple of *current direction*, *current speed* and *current acceleration*.

Strategy The reason why a node has a certain *kinetical state*.

---

[1]Node density of tracked people could be obtained, but not the real node density, since not all tracked people would move in the same area and vice versa there will be lot of people in the area which do not participate in the tracking experiment.

[2]The ability to locate the user of a mobile phone, will also be required by law in some countries. Obvious purpose is to find people quicker in case of an emergency. Of course this could also be used by law enforcement to locate criminals or just to generally watch a country's citizens, which may be desired by totalitarian regimes.

[3]A military drone, that is controlled by a computer, would not fall in this category. However I argue, that for the implications of my assumption it will not matter.

To explain this further, I look again at the existing scenarios and models. *Smooth is Better Than Sharp* [15] is obviously dealing with the *kinetical state* properties in the first place. It controls speed and direction changes. However, there is also a *strategy* defined, which controls how such speed or direction changing events can occur.

In the simple scenarios, there is not much control of the *kinetical state*. Speed is chosen from a fixed interval with a predefined distribution. Direction is directly imposed by the *strategy*. There is no correlating rule between these properties. Acceleration is not taken into account at all. In *Random Waypoint* the *strategy* mainly works by choosing certain destination coordinates by random, and deciding not to move for a certain time, after the destination is reached. It is obvious, that the *strategy* of the simple scenarios is a random strategy.

For a good mobility model, both parts must be combined in a sensible way, although the *strategy* is the more important part, since the *kinetical state* is largely determined by the *strategy*. A more detailed *kinetical state* model (as in [15]), will probably have not as much impact on performance results from simulations, as a more realistic *strategy* (cf. section 4.3.1).

6.3.1 Movement Strategies

A key element, that is missing in most used *strategies* (which act more or less just random), is that nodes do not act just for themselves in the majority, but interact. Typically nodes interact with each other, but also interact with the environment in a matter that affects more than one node at a certain time in a certain region in most cases. The way, how nodes interact with each other and the environment is highly dependent on the role a node assumes. Cars move different than pedestrians and these are much different from paratroopers.

The formation of groups, which is already considered in some mobility models, is a good example for node interaction. Other examples (focused on people in an urban area) include:

- People want to meet each other

- People want to visit a certain location (office building, shop, cinema, conference room).

- Cars form lanes and keep a certain distance to other cars (and other obstacles).

- Traffic lights cause a set of nodes to stop in a relatively small area for a certain time interval.

- Pedestrians will stay on the sidewalk if possible.

- Cars will always stay on the road.

- ...

All these interactions cause a certain concentration of nodes at certain locations. These locations can be considered as hot-spots, since nodes tend to
appear in groups at these locations and thus will increase the node density.
Since node density is a factor of important influence (increased node density
results in increased competition about the physical channel) on the performance of many ad hoc routing protocols, it can be deduced, that this interactive behavior of nodes should be part of the investigation. A pattern mainly
consisting of random movements, may show useless results, if the modeled
scenarios never appear in the real world (or in very rare cases).
Group mobility is an aspect of such behavior that is already part of some
existing models.

6.4   Correlated Movement and Group Mobil-
ity

Group mobility is an important factor, since correlated movement, is likely
to has a significant impact on the communication. Many nodes within a
group, i.e. in close distance to another will have to compete for the physical
channel which causes collisions. Groups are not the only situation, where
nodes will apply a certain correlated movement behavior. Here are some
more examples:

- Pedestrian areas and sidewalks

- Streets

- Traffic lights

These are examples from outdoor traffic. Sidewalks force pedestrians to share
a relatively small space, although the overall space available is much bigger.
Nodes on opposite sidewalks of the street can communicate but the nodes
still concentrate on the edges of the street.
Streets are the only place, where cars can move. The cars move in lanes,
which is also a correlated movement. Traffic situations affect the movement
speed (e.g. during a traffic congestion).

Traffic lights cause a certain set of nodes to stop in their movement and wait at a certain locations. This also results in a higher node density at the traffic lights during the "red" period.

Realistic scenarios like these have not yet been subject to simulation, although there are much more interesting results to expect. Admittedly, it is difficult to model these kind of scenarios for use with available simulation software.

Chapter 7

Classification and Comparison of MANET Routing Protocols

This chapter introduces and characterizes the proposed routing protocols for mobile ad hoc networks. It tries to be comprehensive, but since the development makes rapid progress, and due to time constraints, not all proposed methods could have been considered.

7.1 Choice of Protocols examined

I have chosen to present a very comprehensive overview, including most protocols in the research area of mobile ad hoc networks. This goal is very hard to achieve, not just because of the huge amount of proposed protocols, but also because of the dynamic nature of this research topic; new algorithms are even developed by the time of this writing. Further there is limited information available to certain strategies, such that some protocols cannot be considered in every detail.

Also I focus on a certain type of application. I do not consider Bluetooth[22] or sensor networks. Both are special cases, which require a certain class of algorithms. Instead I concentrate on methods enabling people to communicate with each other using mobile devices, ranging from a cell phone to a car-fitted computer and communications system. Multicast protocols are not examined, too.

Routing protocols considered in this work are: *ABR, ADV, AODV, CBRP, CGSR, CEDAR, DDR, DREAM, DSDV, DSR, DST, FORP, FSLS, FSR, GEDIR, GPSR, GSR, HSR, LANMAR, LAR, LMR, LRR, OLSR, SSA, STAR, Terminode Routing, TBRPF, TORA, WAR, WRP and ZRP.*

I am aware of the following protocols, that are not considered in this paper, due to an aimed deadline for this work: *DST (Dynamic Source Tracing)*, *BEST (Bandwidth Efficient Source Tracing)*, *NSR (Neighborhood Aware Source Routing)*, *SOAR (Source-Tree On-Demand Adaptive Routing)* and *ZHLS (Zone-based Hierarchical Link State Routing)*.

Protocols can be classified and distinguished in many ways. I will also present the most common classes, and protocols which are members of that class. However, these sets are not always disjoint.

For terms used in the following sections, please check the glossary section in appendix A.

7.2 Classification of Ad Hoc Routing Protocols

The large variety of routing protocols reflects the fact that these protocols do implement routing strategies very differently. I categorize the routing protocols into different classes, that represent the key aspect of their strategy. The classes will not be disjoint, as I define several levels of routing strategies. Such work was already done previously in [53], and to some extent in [136]. However at that time, just a small set of routing protocols has been classified. I will present a set of classes which I believe are representative for the different aspects of routing and which may correspond to some of the classes suggested in [53]. Unlike [53] this work does not try to structure disjoint classes into a tree. Although some characteristics are typically dependent upon others, some are not and need to exist in parallel. Some classes will be of an opposing nature, i.e. the protocols can clearly be distinguished between two disjoint classes, e.g. *reactive* and *proactive*. Others will not have a counterpart with characteristics worth pointing out (like *hierarchical* routing protocols, since *non-hierarchical* protocols do not necessarily share distinguished characteristics, apart from not being hierarchical).

There are common characteristics to all routing protocols, due to the nature of their task. Without going in to detail about things like they all try to forward data packets from a source to a destination, it may still be worth to mention that all examined routing protocols show an integrated robustness. All must cope with situations like broken links and nodes becoming suddenly unavailable. Such "exceptional" situations are in fact normal for a MANET routing protocol. However, all routing protocols examined rely on collaboration. Deliberate disruption due to compromised nodes is still a problem for most MANET routing protocols, although meanwhile some derativates (like SAODV[66], not discussed in this work) have been developed to overcome such attacks.

The diagram in figure 7.1 gives an overview about the classes I have decided to use for my characterization. There are no real relations between these classes other than those indicated by arrows. A double arrow indicates an opposing relation, while a single arrow shows a subclass relation.

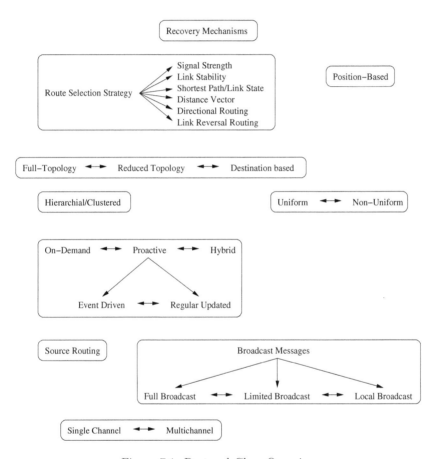

Figure 7.1: Protocol-Class Overview

The choice of some of these classes was inspired by [53].

The reminder of this section will describe each class and which protocols belong to each class. The protocols are denoted with their common abbreviation. The protocols itself are described in detail in chapter 8.

7.2.1 Single Channel vs. Multichannel Protocols

This is essentially a layer 2 property, but several protocols may depend on a certain link layer, while others are specified link-layer independent.

Single channel protocols use just one shared channel to communicate. The IEEE 802.11 DCF medium access method is the most widely used example for such a shared channel link layer. Multichannel protocols utilize CDMA, FDMA or TDMA to form specific channels. Although communication can be much more efficient using such a method, it is difficult to be used in an ad hoc network, since usually a distinguished controlling station is needed to assign the channels.

There are many protocols which do not specify the link layer, but their performance may still depend on it.

Multichannel protocols The following protocols require a multichannel link-layer either explicitly, or their performance depends heavily on it.

CGSR Clusterhead Gateway Switched Routing requires TDMA within a cluster and CDMA between clusters.

TLR/TRR [75] states that the considerations for a link layer protocol for the *Terminode* project center around CDMA.

TORA Implementations of TORA did rely on the encapsulation protocol IMEP [98] used as an underlying secure link layer protocol. IMEP, however did perform very badly together with the IEEE 802.11 Wireless LAN standards and DCF. It was suggested in [40] that other link layer techniques should be used with TORA.

Protocols that use the IEEE 802.11 or a related link layer This class includes all protocols that use a CSMA/CA, MACAW, IEEE 802.11 WLAN with DCF or related link layers. This is the great majority. I just list the protocols, by their abbreviated name:
ADV, AODV, CEDAR, DSR, GPSR, FSLS/HSLS, LANMAR, OLSR.

Unspecified link layer: This remaining list contains all protocols that did not specify a link layer.
ABR, CBRP, DDR, DREAM, DSDV, DST, FORP, FSR, GEDIR, GSR, LAR, LMR, SSA, TBRPF, WAR, WRP.

7.2.2 Uniform vs. Non-Uniform Protocols

As defined in [53], a uniform protocol does not assign any special roles to any node. In a non-uniform protocol some nodes may be assigned a special role, which needs to be performed in a distributed fashion. Typically clustering protocols are non-uniform, as clusterhead nodes are a assigned their special role.

Non-Uniform Protocols Apart from the following *non-uniform* protocols, all others are *uniform*:

CBRP The cluster based routing protocol forms clusters and thus requires clusterheads, which are distinguished nodes.

CGSR The same applies to CGSR, which additionally defines gateway nodes.

CEDAR forms a "core network" (like a backbone), which requires a special role for the nodes, which are part of the core.

DST also creates a backbone on the stable regions of the network.

HSR forms clusters like CBRP and CGSR, but there are no gateway nodes, but multilevel clusters and clusterheads.

LANMAR needs landmark nodes for each group of nodes.

OLSR requires the selection of MPR (multi-point relay) nodes, which is also a special role.

7.2.3 Structure of Topology

The network topology, although dynamic, needs to be structured to be used for routing. Different approaches for structuring are used. I use the main structures of topology for classification. Some of them are closely related to another.
The main topology structures are:

- Flat

- Hierarchical

- Clustered

- Directed Acyclic Graph (DAG)

- Tree

I do a further differentiation by the amount of levels in a hierarchy:

- single level (1)

- two levels (2)

- multiple levels (m)

Hierarchical Topology/Clustered Routing

Clustering is often discussed in the ad hoc networking context. The idea is, to use clusters to introduce some structure into the (otherwise very chaotic and) dynamic nature of the network.

Clusters are usually represented by a dedicated node: the *clusterhead*. This node forms the cluster and attached nodes use the cluster head to describe the cluster they belong to. Clusters can also be formed hierarchically, such that there are multiple layers of clusters. The clusterheads are usually responsible for managing communication within a cluster and are informed about joining and leaving nodes. Additionally to clusterheads, *gateway nodes* are suggested in CGSR and in HSR. These are responsible to transmit information from one cluster to another and therefore may be part of more than one cluster.

Since cluster formation and election of clusterheads is usually a significant effort in terms of signaling traffic, as is the removal and addition of nodes from/to a cluster, cluster-stability has become one important aspect of clustering algorithms.

However, clustering in general does suffer from some drawbacks, especially with very stable clusters. Since the clusterhead and also the gateway nodes have to do the routing and managing work, they can easily become a bottleneck. The communication load will certainly be higher for a clusterhead or a gateway node than for an ordinary node, thus consuming more energy which can lead to an early outage of these nodes due to exhausted power source.

There are also other hierarchical properties I take into account in this class. Some protocols (FSR, DREAM, FSLS) introduce a set of scopes for routing information. In any of these protocols, close, fast moving nodes receive more information more frequently than others. Further there are routing protocols, which use different routing strategies, depending where and how far a packet has traveled, like Terminode Routing or ZRP.

Routing Protocols that use Clustering or a hierarchical Structure:

CBRP The Clustered Routing Protocol defines clusters and clusterheads.

CGSR The Clusterhead Gateway Switch Routing Protocol routes alternating between cluster gateways and clusterheads.

FSLS The Fuzzy Sighted Link State protocol (as well as it's derivates like HSLS), define scopes for dissemination of routing information, and thus also can be considered hierarchical, but not clustered.

FSR The same applies to Fisheye State Routing.

HSR The Hierarchical State Routing protocol does physical and logical clustering. It also is capable of multilevel clustering.

LANMAR Since a landmark can be considered a representative node on a higher level, LANMAR can also be considered a hierarchical routing method.

ZRP The Zone Routing protocol defines a routing zone, this is also some sort of hierarchy.

(All other protocols can be considered as *non-hierarchical*).

7.2.4 Usage of External Services

Some routing protocols may require information through an additional service. The class of routing protocols, which require an external service discussed here are position based protocols.

Position Based Protocols

[113] explains aspects of position based (often also called "location based") routing in detail. Position based routing algorithms claim that no routing tables need to be maintained and thus no overhead due to route discovery and route maintenance is imposed. But they need to obtain position data of their corresponding destinations, either by an internal discovery process, or by an independent position service, which will then impose overhead to maintain the position information (either proactively or on-demand). Several position services are discussed in [113]. Further in this paper, position based routing algorithms are compared in terms of their characteristics and forwarding strategies.

Greedy algorithms like GPSR (cf. 8.16) use either a *most forward within radius* or a *nearest with forward progress* strategy. It is argued, that NFP is of great advantage if the transmission radius/power can be controlled, and additionally has the benefit of reduced channel competition. Recent studies [92] have shown that power control may not improve channel utilization much, because the longer path lengths (in hops) make up for the benefit. Greedy algorithms can route to a local maximum and need a recovery strategy in this case. Among several suggestions, the planar graph traversal methods seem to be the most reasonable.

DREAM and LAR use a flooding approach, but packets are not sent to all neighbors, but only to those in the right direction of the target (i.e. the packet is forwarded to any node within the *request zone*).

Finally some protocols use a hybrid/hierarchical method, like Terminode Routing. For long distances a greedy directional routing method is used. If the packet is close enough, some non-directional mechanism will guide the packet to the destination.

I will now summarize the position based protocols:

DREAM requires an *all-for-all* position service (each node carries a location table for each other node)[113], and requests are forwarded in the right direction.

GEDIR also uses directional routing. To obtain the right direction, the locations of source and target and intermediate nodes must be known. It is not specified how the location information should be determined.

GPSR forwards always to the node closest to the destination within reach, until the target or a local minimum is reached (i.e. there is no other node within range that is closer to the target). Again, it is not specified how location information should be obtained, except for a vague reference to a location database service (cf. section 8.16).

LAR tries to predict the movement of the target node within a time interval to determine a *request zone* to which the data will be broadcast to. Location information from nodes can be piggy-backed on messages, but again it is unclear, how a node is aware of it's position. Also speed and direction are important parameters, and although they could be derived from the positions to some degree, this is not made clear.

Terminode/AGPF Anchored Path Geodesic Packet Forwarding from the Terminode Project (cf. section 8.27) also uses locations to route packets close to their destination. Two methods (FAPD and DRD, explained in

8.27) are proposed to determine the anchored path, but still a general location service like GPS is required.

Other protocols do not use location information.

7.2.5 Topology Updates

The routing and topology information needs to be kept up to date. This is even more important in such a dynamic environment. The way how and when the information is updated is a major characteristic for classification. Routing information is updated:

- Proactively
- On-Demand
- Both (Hybrid)

The update is triggered by:

- An event (i.e. a change in topology is detected)
- A regular time interval

A routing protocol can maintain routing information either on-demand or proactively (at all times). I characterize the protocols accordingly in this section. Further proactive protocols can be divided into protocols that update routing information in regular intervals and protocols that update on certain events. Finally, there are routing protocols that are hybrid and make use of both methods.

On-Demand or Reactive Protocols

A network using an on-demand protocol will not maintain correct routing information on all nodes for all times. Instead, such routing information is obtained *on demand*. If a node wants to transmit a message, and does not have enough routing information to send the message to the destination, the required information has to be obtained (unless the protocol is using directly a flooding approach to deliver the messages). The node needs at least to know the next hop (among its neighbors) for the packet. Although the node could just broadcast the packet to all neighbors this leads to serious congestion in many cases. However, such broadcasts must be used in a route discovery process, since there is no next-hop information available, yet.

Usually this consists of a broadcast message from the originating node, indicating the desired route. Nodes which have the required information will respond to the originating node, which will eventually choose a route from the replies it received. The broadcast may be limited to travel only a few hops first, before a net-wide broadcast will be issued (which would flood the whole network).

Of course, the route request and selection process must be finished, before the message can be sent. This leads to an initial setup delay for messages, if their route is not known to the node. To limit the impact of this delay, most protocols will use a route cache for once established routes. However, the information in this cache will time out, since in a mobile environment, the routes will be invalid after some time.

Clearly, applications that are used over an on-demand routing protocol need to be tolerant for such an initial setup delay.

The advantage of on-demand routing protocols lies in the fact that the wireless channel (a scarce resource) does not need to carry a lot of routing overhead data for routes, that are not even used. This advantage may diminish in certain scenarios where there is a lot of traffic to a large variety of nodes. Thus the scenario will have a very significant impact on the performance. In such a scenario with lots of traffic to many nodes, the route-setup traffic can grow larger than a constant background traffic to maintain correct routing information on each node. Still, if enough capacities would be available, the reduced efficiency (increased overhead) might not affect other performance measures, like throughput or latency.

I also consider some location based protocols as on-demand protocols, since they determine the direction in which to send the packet on demand and some protocols may even initiate a location query of the destination nodes for their packets on demand.

Thus, examples for on-demand protocols are the following:

ABR, AODV, CEDAR, DREAM[1], DSR, FORP, GEDIR, LAR, SSR, WAR.

Proactive Protocols

Proactive routing protocols will try to maintain correct routing information on all nodes in the network at all times. This can be achieved in different ways, and thus divides the protocols into two subclasses: *event driven* and *regular updated* protocols.

---

[1]This algorithm does not initiate a routing selection process. Instead it uses directional routing, but the direction is obtained on demand.

Event driven protocols will not send any routing update packets, if no change in topology occurs. Only if a node detects a change of the topology (usually a change in the neighbor set, or the reception of a message indicating a change in some other nodes neighbor set), this is reported to other nodes, according to the strategy of the routing protocol.

Protocols that are updated in regular intervals will always send their topology information to other nodes at regular intervals. Many link state protocols work in such a manner (but varying the maximum distance of an update message with the length of the interval). Nodes farther away get updates less frequently than close nodes, thus balancing the load imposed on the network.

Proactive protocols of either subclass impose a fixed overhead to maintain the routing tables. Even if many of the entries are not used at all. Their advantage is, that the routes can be used at once and there is no setup delay. [80] compares "flooding protocols"[2] with "hello protocols" (those that periodically announce their neighbors and routes) in terms of overhead in an analytical way.

Event driven proactive routing protocols are the following: CBRP, CGSR, DSDV, GSR, LMR, TORA and WRP.

Regular updated protocols are: DDR, FSLS, FSR, GPSR, LANMAR, OLSR, STAR and TBRPF.

Hybrid Protocols

Also, there are protocols (as to say protocol sets) that utilize both proactive and on-demand routing.

These are:

ADV - Adaptive Distance Vector Routing Routes are maintained proactively, but only to certain nodes (active receivers), and the size and frequency of the updates is adapted. So the authors claim its a hybrid protocol.

Terminode Routing Terminode Routing consists of an on-demand location based component: AGPF (Anchored Path Geodesic Packet Forwarding) and a proactive local routing component (Terminode Local Routing, TLR), which works similar to IARP from ZRP.

---

[2]which is the authors' of [80] term for on-demand routing protocols, which distribute route requests by full broadcast (i.e. flooding)

ZRP - Zone Routing Protocol The Zone Routing Protocol also consists of a proactive Intra Zone Routing Protocol (IARP) and an on-demand Inter Zone Routing Protocol (IERP).

7.2.6 Amount of Topology Information Maintained

Many Routing Protocols transmit topology information, but not all distribute the complete topology information they are aware of. It is difficult to classify the protocols according to this characteristic. Also even if full topology information is maintained in each node, the messages usually only carry sufficient information to reflect the changes in topology but never the whole topology information, since that would not scale.

Full topology is maintained in: DDR, GSR, OLSR, STAR (in ORA mode), TBRPF (in *full topology* mode).

Reduced Topology is maintained in: FSLS, FSR, LANMAR, STAR (in LORA mode), TBRPF (in *partial topology mode*), WRP, ZRP.

This kind of classification is either not applicable to the remaining routing protocols or their role remains uncertain.

7.2.7 Use of Source Routing

A few routing protocols utilize source routing. This means, forwarding depends on the source of the message. Commonly, the source puts all the routing information into the header of a packet. Forwarding nodes utilize this information. In some cases, the forwarding nodes may alter the routing information in the packet to be forwarded. They are just a few protocols using source routing: CBRP, DSR, Terminode/AGPF and WAR.

7.2.8 Use of Broadcast Messages

Broadcast can have different meanings in a wireless environment. There is a *full net-wide broadcast*, which means, a message is intended for every node in the network, and needs to be retransmitted by intermediate nodes. On the other hand, there is a *local broadcast*, which is intended for any node within the senders reach (i.e. the node's neighborhood), but which is not retransmitted at all. In between there are limited broadcasts, which have a limited maximum hop count (time to live) as desired.

There is no routing protocol, that *always* issues full broadcasts, but there are some, that may use full broadcasts: ABR, ADV, AODV, CEDAR, DSDV, DSR, FORP and WAR.

Many protocols prefer a limited broadcast: AODV, FSLS, FSR, HSR, LAN-MAR, LAR, LMR, SSR, Terminode and ZRP.

And also there are protocols, which use only local broadcasts: DDR, GSR, GPSR, OLSR, STAR, TBRPF, TORA and WRP.

Finally, directional routing protocols do not use broadcasts by intention, but would use local multicasts (like a local broadcast, but not addressed to all neighbors), like DREAM and GEDIR.

7.2.9 Recovery Mechanisms

Since the routing information in each node may become stale, some protocols may need a route recovery or route salvage mechanism. It is clear, that proactive routing protocols do not need a specific recovery mechanism, since they react to topology changes anyway within a short period. On-Demand protocols, however, need to fix routes which are not available any more.

The following protocols have some (explicit or implicit) recovery mechanism: ABR, AODV, CBRP, DREAM[3] DSR, FORP, WAR and ZRP.

The following protocols could utilize such a mechanism, but do not support one: ADV, GEDIR, LAR.

7.2.10 Route Selection Strategy

The route selection strategy is an important aspect of a routing protocol. I describe the main representatives and the protocols, which use them.

Signal Strength: Route packets along the connection with the best signal strength. This is used by ABR and SSR.

Link Stability: Route packets along the connections that appear most stable over a period of time. It is is used by DST and FORP

Shortest Path/Link State: Select a shortest path according to some metric. This is used by many protocols: CEDAR, DDR, FSR, GSR, HSR, LANMAR, OLSR, STAR, TBRPF.

---

[3]The recovery mechanism is not specified, just a `Recovery()` routine is mentioned.

Distance Vector: The common distance vector method, usual by hop count, is used by ADV, AODV, DSDV, DSR, WRP, ZRP.

Directional Routing: This routes into the geographic direction of the target and is mainly used by location based protocols: DREAM, GEDIR, GPSR, LAR, Terminode/AGPF.

Link Reversal Routing: is a routing family which is used by LMR and TORA (cf. also section 8.22). It is based on flows in a graph.

7.3 Possible Dependencies Between Protocol Characteristics

In this section, I examine possible dependencies between the chosen characteristics. The aim was to provide orthogonal characteristics with as little dependencies as possible. Since this is not always perfectly feasible, I explain the possible dependencies.

Channel Usage might influence *route selection*, where *signal strength* or *link stability* are used as criteria and *uniformity*, since a multi-channel method might need special nodes for coordination. However, the choice does not seem to restrict any other parameter.

Uniformity might influence *route selection*, where clusters or special nodes (like Multi-point Relays in OLSR 8.23) are used. It might be influenced by *channel usage*, to provide special nodes to coordinate channel multiplexing.

External Information will influence *route selection*, since external services will be required for *position based routing* and possibly also for *signal strength routing*.

Source Routing will influence *topology information*, since the determination of the complete route in the source requires accurate and complete topology information in each node. Further it helps with some *recovery mechanisms* as described in WAR, section 8.29.

Accuracy/Update Frequency will influence *topology updates* which use *full broadcasts*, since frequent updates with full broadcasts will lead to congestion. Also maintenance of *full topology* might be influenced, since this would lead to lots of updates. Further *route selection* methods could be influenced by this parameter.

Route discovery/Obtain Topology/Use of Broadcasts will influence *route selection*, since many selection strategies will require some route discovery mechanisms. Also the maintained topology needs to be updated by some discovery/maintenance mechanism.

Route Selection will in turn influence many other characteristics like *channel usage, uniformity*, may require *external information*. Especially if certain methods like positional routing or clustered routing is used.

7.4 Comparison Functions for Routing Protocol Characteristics

This section describes the comparison functions for each characteristic in more detail:

7.4.1 Overview over Comparison Function

As described in section 3.3.1, a sequence code must be established to represent the characteristics of the routing protocols. A set of functions can be defined on the characteristics represented in the code to compare the protocols with each other and yield an affinity value just like the simulation scenarios and the applications. Sometimes there are more than one parameter to represent a characteristic and thus more complex functions using combinations of parameters are used. Please note that not always all possible combinations are encountered (and would make sense), like in the example of **Topology Structure** below.

In the following, the function for each characteristic is described and explained in detail.

If for any characteristic a value cannot be determined, a "?" is used instead, which usually yields a 0.5 matching value (cf. section 4.7.1).

7.4.2 Description of Comparison Functions

Channel Use Compares how the protocols uses the wireless channel, i.e. if a *single channel, "S"* is used or if *multiple channels, "M"* are used. If there is a match, 1.0 is returned, but if there is no match, there is still a result of 0.2, as even single channel protocols may use some sort of multiplexing mechanism.

This characteristic has a weight of 3.

$$f_1 : \{M, S, ?\} \times \{M, S, ?\} \mapsto [0 : 1]$$

$f_1(\gamma_{ch}, \gamma'_{ch})$	M	S	?
M	1.0	0.2	0.5
S	0.2	1.0	0.5
?	0.5	0.5	0.25

Diversity of Nodes This compares, if all nodes have the same roles in the network (*uniform*, "*U*") using this protocol, or if there are nodes with a special role (*non-uniform*, "*N*"). If there is a match, 1.0 is returned, and if there is no match, 0.1 is returned, since in the non-uniform protocol, the nodes might still be very similar or just very few with special roles.

This characteristic has a weight of 3.

$$f_2 : \{N, U, ?\} \times \{N, U, ?\} \mapsto [0 : 1]$$

$f_2(\gamma_{div}, \gamma'_{div})$	N	U	?
N	1.0	0.1	0.5
U	0.1	1.0	0.5
?	0.5	0.5	0.25

Use of Source Routing Here are two aspects compared. First, if the routing decision is taken at the source ($\rho_d = S$) or in each router ($\rho_d = R$) and second if the route is stored in the packet header ($\rho_s = P$) or in a routing table ($\rho_s = T$). This is a straigntforward match and 1.0 are only returned if both aspects are matching.

This characteristic has a weight of 4.

$$f_3 : \{S, R, ?\} \times \{S, R, ?\} \times \{P, T, ?\} \times \{P, T, ?\} \mapsto [0 : 1]$$
$$f_3(\rho_d, \rho_s, \rho'_d, \rho'_s) = \frac{1}{2}(g_3(\rho_d, \rho'_d) + h_3(\rho_s, \rho'_s))$$

$$g_3 : \{S, R, ?\} \times \{S, R, ?\} \mapsto [0 : 1]$$

$g_3(\rho_d, \rho'_d)$	S	R	?
S	1.0	0	0.5
R	0	1.0	0.5
?	0.5	0.5	0.25

$$h_3 : \{P, T, ?\} \times \{P, T, ?\} \mapsto [0 : 1]$$

$h_3(\rho_s, \rho_s')$	P	T	?
P	1.0	0	0.5
T	0	1.0	0.5
?	0.5	0.5	0.25

Use of External Information It is compared if external information is required for the routing and what type of information. Possible values for required information (ι_i) are *Location*, "*L*", *Signal Strength*, "*S*" or *none*, "*N*". Additionally it is compared if any special requirements are needed for this information. Possible values for the requirements (ι_r) are an additional *service*, "*S*" or special *hardware*, "*H*" is required.

If the type of information is matching, it yields 1.0 for this subcomparison and 0 if there is no match. If there are special requirements for an additional service for both protocols ($\iota_r = \iota_r' = S$) a maximum of 0.8 is added for the second subcomparison. Only 0.8 are chosen, since the type of additional service and requirements are unknown and could differ a lot. If the both protocols require special hardware ($\iota_r = \iota_r' = H$) the result is even less (0.2), because special hardware can vary a great deal (from directional antennas to GPS receivers).

This characteristic has a weight of 4.

$$f_4 : \{N, L, S, ?\} \times \{N, L, S, ?\} \times \{N, S, H, ?\} \times \{N, S, H, ?\} \mapsto [0:1]$$

$$f_4(\iota_i, \iota_r, \iota_i', \iota_r') = \frac{1}{2}(g_4(\iota_i, \iota_i') + h_4(\iota_r, \iota_r'))$$

$$g_4 : \{N, L, S, ?\} \times \{N, L, S, ?\} \mapsto [0:1]$$

$g_4(\iota_i, \iota_i')$	N	L	S	?
N	1.0	0	0	0.5
L	0	1.0	0	0.5
S	0	0	1.0	0.5
?	0.5	0.5	0.5	0.25

$$h_4 : \{N, S, H, ?\} \times \{N, S, H, ?\} \mapsto [0:1]$$

$h_4(\iota_r, \iota_r')$	N	S	H	?
N	1.0	0	0	0.5
S	0	0.8	0	0.5
H	0	0	0.2	0.5
?	0.5	0.5	0.5	0.25

Routing Information Update This compares how the routing information is updated and on what occasion. Again two subcomparisons are
used. Possible values for the way of updates (v_u) are *On-Demand, "O",
Proactive, "P"* and *Hybrid, "H"*. The update schedule (v_o) comparison
allows the values of *Event Driven, "E", Regular interval driven, "R"* or
Both, "B". The comparison of the type has double weight within the
comparison. If both types are equal thus 1.0 is yielded for this subcomparison (counting double). 0.5 is yielded for any non-matching
combination with a *Hybrid* value. The second comparison can only
yield 1.0 or 0.

This characteristic has a weight of 6.

$$f_5 : \{O, P, H, ?\} \times \{O, P, H, ?\} \times \{E, R, B, ?\} \times \{E, R, B, ?\} \mapsto [0 : 1]$$

$$f_5(v_u, v_o, v_u', v_o') = \frac{1}{3}(2g_5(v_u, v_u') + h_5(v_o, v_o'))$$

$$g_5 : \{O, P, H, ?\} \times \{O, P, H, ?\} \mapsto [0 : 1]$$

$g_5(v_u, v_u')$	O	P	H	?
O	1.0	0	0.5	0.5
P	0	1.0	0.5	0.5
H	0.5	0.5	1.0	0.5
?	0.5	0.5	0.5	0.25

$$h_5 : \{E, R, B, ?\} \times \{E, R, B, ?\} \mapsto [0 : 1]$$

$h_5(v_o, v_o')$	E	R	B	?
E	1.0	0	0	0.5
R	0	1.0	0	0.5
B	0	0	1.0	0.5
?	0.5	0.5	0.5	0.25

Broadcast Usage and Routing Information This characteristic has three
subcomparisons. First if *Full broadcasts* may be used $(\tau_f = F)$ or may
NOT be used $(\tau_f = N)$ (subweight 2). Second if *Limited or Restricted*
broadcasts are used $(\tau_r = R)$, if *Local* broadcasts are used $(\tau_r = L)$ or
if no broadcasts are used at all $(\tau_r = N)$ (again with a subweight of
2) and finally if *external services* are also used $(\tau_e = E)$ to get routing
information or not $(\tau_e = N)$ (weight 1).

The matches yield straightforward values (1.0 on match times weight, 0 on non-match). There is one exception. If the second field does not match (i.e. one protocol uses restricted broadcasts while the other uses only local broadcasts) still a value of 0.3 times weight is added.

This characteristic has an overall weight of 5.

$$f_6 : \{F, N, ?\} \times \{F, N, ?\} \times \{R, L, N, ?\} \times \{R, L, N, ?\} \times \{E, N, ?\} \times \{E, N, ?\} \mapsto [0 : 1]$$

$$f_6(\tau_f, \tau_r, \tau_e, \tau_f', \tau_r', \tau_e') = \frac{1}{5}(2g_6(\tau_f, \tau_f') + 2h_6(\tau_r, \tau_r') + j_6(\tau_e, \tau_e'))$$

$$g_6 : \{F, N, ?\} \times \{F, N, ?\} \mapsto [0 : 1]$$

$g_6(\tau_f, \tau_f')$	F	N	?
F	1.0	0	0.5
N	0	1.0	0.5
?	0.5	0.5	0.25

$$h_6 : \{R, L, N, ?\} \times \{R, L, N, ?\} \mapsto [0 : 1]$$

$h_6(\tau_r, \tau_r')$	R	L	N	?
R	1.0	0.3	0	0.5
L	0.3	1.0	0	0.5
N	0	0	1.0	0.5
?	0.5	0.5	0.5	0.25

$$j_6 : \{E, N, ?\} \times \{E, N, ?\} \mapsto [0 : 1]$$

$j_6(\tau_e, \tau_e')$	E	N	?
E	1.0	0	0.5
N	0	1.0	0.5
?	0.5	0.5	0.25

Topology Information Maintained Compares the amount of topology information and state maintained by the routing protocol. Possible values are *Full Topology* ($\theta_i = F$), *Reduced Topology* ($\theta_i = R$), *Full or Reduced (Both)* ($\theta_i = B$) and *Next Hop only* ($\theta_i = N$).

Apart from the standard result of 1.0 for a match there are also the following results for "close" matches: A combination of *Both* and *Next Hop* yields 0.5. A combination of *Full* and *Reduced* topology will yield a combination of 0.3. And a combination of *Reduced Topology* and *Next Hop* yields 0.2.

This characteristic has an overall weight of 5.

$$f_7 : \{F, R, B, N, ?\} \times \{F, R, B, N, ?\} \mapsto [0:1]$$

$f_7(\theta_i, \theta_i')$	F	R	B	N	?
F	1.0	0.3	0.5	0	0.5
R	0.3	1.0	0.5	0	0.5
B	0.5	0.5	1.0	0	0.5
N	0	0	0	1.0	0.5
?	0.5	0.5	0.5	0.5	0.25

Topology Structure This comparison has again two subcomparisons. First the actual topology structure is compared and can have the following values *flat* ($\theta_s = F$), *clustered* ($\theta_s = C$), *hierarchical* ($\theta_s = H$), *tree* ($\theta_s = T$) and *directly acyclic graph* ($\theta_s = D$). The second part compares the amount of structural levels (applicable in hierarchical and clustered topologies) with possible values of *1* ($\theta_l = 1$), *2* ($\theta_l = 2$) or *more* ($\theta_l = m$) levels. The structure counts 3 times while the amount of levels only counts 2 times. Please note, that not all combinations are possible, since a *flat* ($\theta_s = F$) structure implies a single structural level, it will always come with $\theta_l = 1$.

The following combinations of structures yield the followin values: a combination of equal values yields 1.0 (times 3). A combination of *clustered* and *hierarchical* yields 0.5, *clustered* and *tree* yields 0.3 and *tree* and *directly acyclic graph* yields again 0.5.

The amount of levels compare as follows: equal values yield 1.0. *1* and *2* yield 0.6 and *2* and *more* yields 0.4.

This characteristic has an overall weight of 5.

$$f_8 : \{F, C, H, T, D, ?\} \times \{F, C, H, T, D, ?\} \times \{1, 2, m, ?\} \times \{1, 2, m, ?\} \mapsto [0:1]$$

$$f_8(\theta_s, \theta_l, \theta_s', \theta_l') = \frac{1}{5}(3g_8(\theta_s, \theta_s') + 2h_8(\theta_l, \theta_l'))$$

$$g_8 : \{F, C, H, T, D, ?\} \times \{F, C, H, T, D, ?\} \mapsto [0:1]$$

$g_8(\theta_s, \theta'_s)$	F	C	H	T	D	?
F	1.0	0	0	0	0	0.5
C	0	1.0	0.5	0.3	0	0.5
H	0	0.5	1.0	0.3	0	0.5
T	0	0.3	0.3	1.0	0.5	0.5
D	0	0	0	0.5	1.0	0.5
?	0.5	0.5	0.5	0.5	0.5	0.25

$$h_8 : \{1, 2, m, ?\} \times \{1, 2, m, ?\} \mapsto [0 : 1]$$

$g_8(\theta_s, \theta'_s)$	1	2	m	?
1	1.0	0.3	0	0.5
2	0.3	1.0	0.2	0.5
m	0	0.2	1.0	0.5
?	0.5	0.5	0.5	0.25

Route Selection Criteria This compares how a route is selected. Possible values are *Distance Vector* ($\sigma = V$), *Link State/Shortest Path* ($\sigma = L$), *Signal Strength* ($\sigma = S$), *Direction* ($\sigma = D$), *Link Stability* ($\sigma = B$), *Flow Algorithm* ($\sigma = A$), *Cluster* ($\sigma = C$). No special combination values are used.

This characteristic has a weight of 6.

$$f_9 : \{V, L, S, D, B, A, C, ?\} \times \{V, L, S, D, B, A, C, ?\} \mapsto [0 : 1]$$

$f_9(\sigma, \sigma')$	V	L	S	D	B	A	C	?
V	1.0	0	0	0	0	0	0	0.5
L	0	1.0	0	0	0	0	0	0.5
S	0	0	1.0	0	0	0	0	0.5
D	0	0	0	1.0	0	0	0	0.5
B	0	0	0	0	1.0	0	0	0.5
A	0	0	0	0	0	1.0	0	0.5
C	0	0	0	0	0	0	1.0	0.5
?	0.5	0.5	0.5	0.5	0.5	0.5	0.5	0.25

Recovery Capabilities This compares if the routing protocol takes any special precautions to recover from possible errors. Possible values are *no capabilities* ($\chi = N$), *recovery extensions possible* (but not implemented) ($\chi = P$) and *packet salvage* (one-hop packet retransmit) ($\chi = S$). Again no special values for certain combinations are used.

The weight of this characteristic is 3.

$$f_{10} : \{N, P, S, ?\} \times \{N, P, S, ?\} \mapsto [0 : 1]$$

$f_{10}(\chi, \chi')$	N	P	S	?
N	1.0	0	0	0.5
P	0	1.0	0	0.5
S	0	0	1.0	0.5
?	0.5	0.5	0.5	0.25

These functions are also each associated with a weight. The weights have been chosen to emphasise the methods to update and maintain routing information and the way routes are selected.

Consequently, the final result is computed as follows: $f = \frac{1}{44}(3f_1 + 3f_2 + 4f_3 + 4f_4 + 6f_5 + 5f_6 + 5f_7 + 5f_8 + 6f_9 + 3f_{10})$

Example

The following examples will illustrate, how the comparison works. In the first example (table 7.1) AODV (cf. section 8.3) will be compared with DSR (cf. section 8.10). In the second example (table 7.2) AODV will be compared with STAR (cf. section 8.25).

Characteristic	Value AODV	Value DSR	W	Result
Channel Use	$\gamma_{ch} = S$	$\gamma'_{ch} = S$	3	3.0
Diversity of Nodes	$\gamma_{div} = U$	$\gamma'_{div} = U$	3	3.0
Use of Source Routing	$\rho_d = R$	$\rho'_d = S$	4	0.0
	$\rho_s = T$	$\rho'_s = P$		
Use of External Information	$\iota_i = N$	$\iota'_i = N$	4	4.0
	$\iota_r = N$	$\iota'_r = N$		
Routing Information Update	$\upsilon_u = O$	$\upsilon'_u = O$	6	6.0
	$\upsilon_o = E$	$\upsilon'_o = E$		
Broadcast Usage and	$\tau_f = F$	$\tau'_f = F$	6	5.0
Routing Information	$\tau_r = R$	$\tau'_r = R$		
	$\tau_e = N$	$\tau'_e = N$		
Topology Information Maintained	$\theta_i = N$	$\theta'_i = N$	5	5.0
Topology Structure	$\theta_s = F$	$\theta'_s = F$	5	5.0
	$\theta_l = 1$	$\theta'_l = 1$		
Route Selection Criteria	$\sigma = V$	$\sigma' = V$	6	6.0
Recovery Capabilities	$\chi = P$	$\chi' = P$	3	3.0
Sum:			44	40.0
Result				0.91

Table 7.1: Similarity Comparison of AODV and DSR

Characteristic	Value AODV	Value STAR	W	Result
Channel Use	$\gamma_{ch} = S$	$\gamma'_{ch} = ?$	3	1.5
Diversity of Nodes	$\gamma_{div} = U$	$\gamma'_{div} = U$	3	3.0
Use of Source Routing	$\rho_d = R$	$\rho'_d = R$	4	4.0
	$\rho_s = T$	$\rho'_s = T$		
Use of External Information	$\iota_i = N$	$\iota'_i = N$	4	4.0
	$\iota_r = N$	$\iota'_r = N$		
Routing Information Update	$v_u = O$	$v'_u = P$	6	0.0
	$v_o = E$	$v'_o = R$		
Broadcast Usage and	$\tau_f = F$	$\tau'_f = N$	5	1.6
Routing Information	$\tau_r = R$	$\tau'_r = L$		
	$\tau_e = N$	$\tau'_e = N$		
Topology Information Maintained	$\theta_i = N$	$\theta'_i = B$	5	0.0
Topology Structure	$\theta_s = F$	$\theta'_s = T$	5	0.0
	$\theta_l = 1$	$\theta'_l = m$		
Route Selection Criteria	$\sigma = V$	$\sigma' = L$	6	0.0
Recovery Capabilities	$\chi = P$	$\chi' = N$	3	0.0
Sum:			44	14.1
Result				0.32

Table 7.2: Similarity Comparison of AODV and STAR

7.4.3 Comparison Results

The following tables and figures 7.2,7.3 and 7.4 show a visualization of the similarity-relationships of each protocol, with a similarity of 60%, 70% and 80% correspondingly. The colors of the numbers and edges denote the degree of similarity [4].

Color	Similarity s in %
red	$s \geq 90$
orange	$80 \leq s < 90$
light green	$70 \leq s < 80$
dark green	$60 \leq s < 70$
blue	$50 \leq s < 60$
black	$0 \leq s < 50$

---

[4]Please note that in the colored value matrix some border values (e.g. 70%) may appear in different colors, due to rounding.

Proto	ABR	ADV	AODV	CBRP	CEDAR	CGSR	DDR	DREAM
ABR	–	0.67	0.83	0.40	0.60	0.44	0.30	0.65
ADV	0.67	–	0.84	0.35	0.61	0.49	0.43	0.51
AODV	0.83	0.84	–	0.42	0.63	0.42	0.32	0.67
CBRP	0.40	0.35	0.42	–	0.45		0.33	0.29
CEDAR	0.60	0.61	0.63	0.45	–	0.59	0.49	0.44
CGSR	0.44	0.49	0.42		0.59	–	0.50	0.33
DDR	0.30	0.43	0.32	0.33	0.49	0.50	–	0.22
DREAM	0.65	0.51	0.67	0.29	0.44	0.33	0.22	–
DSDV	0.65	0.88	0.81	0.42	0.58	0.60	0.46	0.49
DSR			0.91	0.51	0.54	0.33	0.23	0.58
DST	0.51	0.59	0.52	0.39	0.56	0.57	0.57	0.39
FORP	0.81	0.67	0.83	0.40	0.60	0.44	0.30	0.65
FSLS	0.43	0.58	0.47	0.36	0.61	0.50		0.32
FSR	0.36	0.49	0.38	0.36	0.60	0.54		0.24
GEDIR	0.59	0.58	0.60	0.22	0.51	0.39	0.29	0.80
GPSR	0.33	0.48	0.37	0.35	0.47	0.48	0.52	0.61
GSR	0.41	0.49	0.42	0.36	0.57	0.54	0.85	0.32
HSR	0.47	0.46	0.48	0.41	0.83	0.59	0.61	0.35
LANMAR	0.39	0.54	0.42	0.35	0.67	0.48		0.28
LAR	0.65	0.51	0.67	0.31	0.44	0.35	0.24	0.89
LMR	0.49	0.58	0.51	0.47	0.53	0.64	0.59	0.38
OLSR	0.27	0.42	0.31	0.38	0.67	0.52	0.82	0.19
SSA		0.60	0.62	0.24	0.53	0.42	0.33	0.59
STAR	0.30	0.43	0.32	0.33	0.49	0.50	0.89	0.22
TBRPF	0.30	0.43	0.32	0.33	0.49	0.50	0.89	0.22
TORA	0.46	0.55	0.48	0.47	0.50	0.64	0.62	0.38
Terminode	0.44	0.58	0.42	0.35	0.33	0.36	0.30	0.64
WAR	0.65		0.81	0.42	0.51	0.35	0.21	0.49
WRP	0.49		0.64	0.33	0.41	0.51	0.57	0.40
ZRP	0.49	0.66	0.64	0.35	0.47	0.39	0.46	0.37

Proto	DSDV	DSR	DST	FORP	FSLS	FSR	GEDIR	GPSR
ABR	0.65	0.71	0.51	0.81	0.43	0.36	0.59	0.33
ADV	0.88	0.75	0.59	0.67	0.58	0.49	0.58	0.48
AODV	0.81	0.91	0.52	0.83	0.47	0.38	0.60	0.37
CBRP	0.42	0.51	0.39	0.40	0.36	0.36	0.22	0.35
CEDAR	0.58	0.54	0.56	0.60	0.61	0.60	0.51	0.47
CGSR	0.60	0.33	0.57	0.44	0.50	0.54	0.39	0.48
DDR	0.46	0.23	0.57	0.30	0.80	0.79	0.29	0.52
DREAM	0.49	0.58	0.39	0.65	0.32	0.24	0.80	0.61
DSDV	–	0.71	0.57	0.65	0.59	0.52	0.56	0.49
DSR	0.71	–	0.43	0.71	0.38	0.28	0.51	0.28
DST	0.57	0.43	–	0.64	0.56	0.55	0.46	0.53
FORP	0.65	0.71	0.64	–	0.43	0.36	0.59	0.33
FSLS	0.59	0.38	0.56	0.43	–	0.91	0.39	0.57
FSR	0.52	0.28	0.55	0.36	0.91	–	0.31	0.56
GEDIR	0.56	0.51	0.46	0.59	0.39	0.31	–	0.68
GPSR	0.49	0.28	0.53	0.33	0.57	0.56	0.68	–
GSR	0.56	0.33	0.52	0.41	0.42	0.44	0.39	0.50
HSR	0.44	0.39	0.52	0.47	0.65	0.69	0.42	0.39
LANMAR	0.55	0.33	0.53	0.39	0.94	0.93	0.34	0.63
LAR	0.49	0.58	0.41	0.65	0.36	0.29	0.83	0.62
LMR	0.65	0.42	0.65	0.49	0.64	0.61	0.45	0.53
OLSR	0.43	0.22	0.45	0.27			0.26	0.57
SSA	0.59	0.53	0.48	0.61	0.45	0.38	0.65	0.45
STAR	0.46	0.23	0.57	0.30	0.82	0.81	0.29	0.52
TBRPF	0.46	0.23	0.57	0.30	0.82	0.81	0.29	0.52
TORA	0.62	0.39	0.65	0.46	0.60	0.58	0.45	0.57
Terminode	0.51	0.51	0.41	0.44	0.39	0.35	0.71	0.60
WAR	0.70	0.90	0.41	0.65	0.34	0.27	0.49	0.24
WRP		0.55	0.48	0.49	0.70	0.62	0.47	0.48
ZRP	0.55	0.55	0.44	0.49	0.61	0.62	0.30	0.43

Proto	GSR	HSR	LANMAR	LAR	LMR	OLSR	SSA	STAR
ABR	0.41	0.47	0.39	0.65	0.49	0.27		0.30
ADV	0.49	0.46	0.54	0.51	0.58	0.42	0.60	0.43
AODV	0.42	0.48	0.42	0.67	0.51	0.31	0.62	0.32
CBRP	0.36	0.41	0.35	0.31	0.47	0.38	0.24	0.33
CEDAR	0.57	0.83	0.67	0.44	0.53	0.67	0.53	0.49
CGSR	0.54	0.59	0.48	0.35	0.64	0.52	0.42	0.50
DDR	0.85	0.61		0.24	0.59	0.82	0.33	0.89
DREAM	0.32	0.35	0.28	0.89	0.38	0.19	0.59	0.22
DSDV	0.56	0.44	0.55	0.49	0.65	0.43	0.59	0.46
DSR	0.33	0.39	0.33	0.58	0.42	0.22	0.53	0.23
DST	0.52	0.52	0.53	0.41	0.65	0.45	0.48	0.57
FORP	0.41	0.47	0.39	0.65	0.49	0.27	0.61	0.30
FSLS		0.65	0.94	0.36	0.64		0.45	0.82
FSR		0.69	0.93	0.29	0.61		0.38	0.81
GEDIR	0.39	0.42	0.34	0.83	0.45	0.26	0.65	0.29
GPSR	0.50	0.39	0.63	0.62	0.53	0.57	0.45	0.52
GSR	–	0.65		0.34	0.61	0.80	0.43	
HSR	0.65	–	0.68	0.40	0.54		0.49	0.61
LANMAR		0.68	–	0.32	0.59	0.83	0.41	
LAR	0.34	0.40	0.32	–	0.43	0.20	0.63	0.24
LMR	0.61	0.54	0.59	0.43	–	0.47	0.52	0.59
OLSR	0.80		0.83	0.20	0.47	–	0.30	
SSA	0.43	0.49	0.41	0.63	0.52	0.30	–	0.33
STAR		0.61		0.24	0.59		0.33	–
TBRPF		0.61		0.24	0.59		0.33	0.95
TORA	0.64	0.50	0.56	0.39	0.92	0.50	0.49	0.62
Terminode	0.35	0.33	0.35	0.68	0.44	0.23	0.56	0.30
WAR	0.31	0.38	0.30	0.49	0.40	0.18	0.52	0.21
WRP	0.68	0.46	0.65	0.42	0.58	0.54	0.51	0.60
ZRP	0.49	0.50	0.67	0.42	0.46	0.50	0.37	0.48

Proto	TBRPF	TORA	Terminode	WAR	WRP	ZRP
ABR	0.30	0.46	0.44	0.65	0.49	0.49
ADV	0.43	0.55	0.58	0.72	0.71	0.66
AODV	0.32	0.48	0.42	0.81	0.64	0.64
CBRP	0.33	0.47	0.35	0.42	0.33	0.35
CEDAR	0.49	0.50	0.33	0.51	0.41	0.47
CGSR	0.50	0.64	0.36	0.35	0.51	0.39
DDR	0.89	0.62	0.30	0.21	0.57	0.46
DREAM	0.22	0.38	0.64	0.49	0.40	0.37
DSDV	0.46	0.62	0.51	0.70	0.78	0.55
DSR	0.23	0.39	0.51	0.90	0.55	0.55
DST	0.57	0.65	0.41	0.41	0.48	0.44
FORP	0.30	0.46	0.44	0.65	0.49	0.49
FSLS	0.82	0.60	0.39	0.34	0.70	0.61
FSR	0.81	0.58	0.35	0.27	0.62	0.62
GEDIR	0.29	0.45	0.71	0.49	0.47	0.30
GPSR	0.52	0.57	0.60	0.24	0.48	0.43
GSR	0.79	0.64	0.35	0.31	0.68	0.49
HSR	0.61	0.50	0.33	0.38	0.46	0.50
LANMAR	0.77	0.56	0.35	0.30	0.65	0.67
LAR	0.24	0.39	0.68	0.49	0.42	0.42
LMR	0.59	0.92	0.44	0.40	0.58	0.46
OLSR	0.26	0.50	0.23	0.18	0.54	0.50
SSA	0.33	0.49	0.56	0.52	0.51	0.37
STAR	0.95	0.62	0.30	0.21	0.60	0.48
TBRPF	–	0.62	0.30	0.21	0.60	0.48
TORA	0.62	–	0.41	0.37	0.61	0.43
Terminode	0.30	0.41	–	0.53	0.43	0.39
WAR	0.21	0.37	0.53	–	0.53	0.46
WRP	0.60	0.61	0.43	0.53	–	0.66
ZRP	0.48	0.43	0.39	0.46	0.66	–

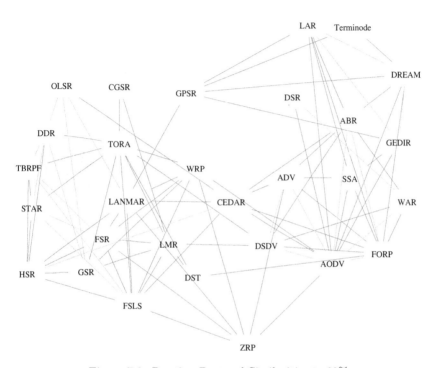

Figure 7.2: Routing Protocol Similarities $\geq 60\%$

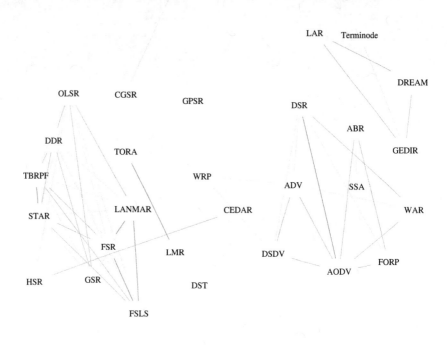

Figure 7.3: Routing Protocol Similarities $\geq 70\%$

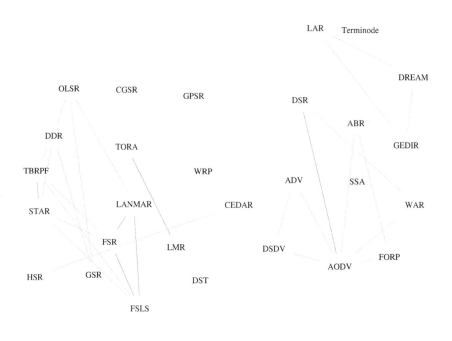

Figure 7.4: Routing Protocol Similarities $\geq 80\%$

This shows which protocols are closely related to another, and which would possibly behave overall in a similar way.

Three main clusters of similar protocols can be identified. First a big cluster of link state like protocols with no clear cluster centre, then a cluster of reactive distance vector protocols around AODV and third, a cluster of geographically oriented protocols, again with no clear cluster centre.

This result could be used to compare these three big classes of routing protocols against a set of benchmark application scenarios using a small set of represeative protocols for each class, e.g. STAR, OLSR and FSR for the link state routing protocols, AODV and DSR for the reactive protocols and LAR for the geographic routing protocols.

This also confirms what is already known by development history of some protocols. Protocols which originated in the same development group or which are commonly oriented on the same principles (like geographic location or link state properties) are grouped together indicating their similarity.

Additionally, each section in the following chapter 8 (which described each protocol in detail) includes a diagram, that shows the degree of similarity to the other protocols. The individual similarities are not commented in detail, since the big picture is already described here.

If protocols with similar design properties do also behave in a similar way (with respect to the performance metrics described in section 2.1.1) is still to be confirmed, i.e. if my choice of functions to describe the similarity are appropriate to yield similar behavior in various application scenarios. This can be done by simulation. See chapter 10 for a more detailed description about the possibilities. However, such a result would then show if some class of routing protocols has certain advantages in one or more particular application environments and thus give valuable input for further research.

7.5 Performance Comparisons Previously Done

In literature many performance comparisons have been done, but not as much, as one could probably expect. Of course it would not be possible or even useful to compare every single routing protocol with every other protocol in any kind of scenario.

The comparisons done so far are summarized in the following sections, including the references and a brief result of the comparison. These comparisons are as well included and discussed in more detailed in the individual protocol descriptions in chapter 8.

7.5.1 ABR vs. DSR and DBF

In [100] ABR was compared to DSR and DBF[5]. The results are in favor of ABR, in terms of overhead, throughput and end-to-end delay, although the advantage to DSR is quite small. Other criteria, like memory requirements for the table and power consumption show disadvantages of ABR.

7.5.2 ADV vs. AODV, DSDV and DSR

In [23] ADV was compared against DSDV, AODV and DSR. The results are clearly in favour of ADV, but the papers is from the same authors as ADV. In [49] ADV was compared against AODV and DSR focusing on TCP traffic and using certain modifications to TCP to improve performance in MANETS (fixed RTO, SACK, delayed ACK). The paper is again well in favour of ADV showing that ADV performs well even without such modifications.

7.5.3 AODV, DSDV, DSR and TORA

These protocols have been compared with each other in [25], [44], [85] and [23]. The results are mixed, but the tendency is that DSDV does not perform so well compared to other (mainly more modern) protocols, while DSR and AODV show better results. TORA did perform worst, but the results have been challenged, due to the possibly unfair implementation of the comparison.

7.5.4 FSR, HSR and DSDV

[78] compares equally all three protocols HSR, FSR and DSDV. While HSR and FSR scale better to high numbers of nodes than DSDV, they are more complex and may suffer from an increased delay. The paper does not explicity favour one of the protocols but observes FSR and HSR may be particularly useful in scenarios where scalability is more important.

Although [153] compares FSR with nine other MANET routing protocols, no performance comparison was done, just a plain comparison of some characteristics.

---

[5]Distributed Bellmann-Ford

7.5.5 GEDIR vs. DIR and MFR

GEDIR was compared to DIR[6] and MFR[7] (both very simple algorithms, possibly only used as a model for the sake of the comparison) in [104]. The paper is in favour of GEDIR, but all three protocols are proposed by the same author of the paper.

7.5.6 GPSR vs. DSR

GPSR was compared to DSR in [89], the introducing paper for GPSR, being in favour of GPSR for the examined scenario. GPSR was also compared to GEAR (not further considered in this book), a geographic energy conserving protocol in [168]. That paper (proposing GEAR) is in favour of GEAR.

7.5.7 GSR vs. DBF and ILS

GSR was compared to DBF[8] and ILS[9] in [32] showing GSR being more accurate than DBF and using less overhead than a traditional (though ideal) link state protocol.

7.5.8 LANMAR vs. AODV, DSR and FSR

[123] proposes LANMAR (an extension to FSR) and compares it against AODV, DSR and plain FSR, being clearly in favour of LANMAR.

7.5.9 OLSR vs. AODV and DSR

[79] does an analytic comparison of OLSR with DSR which gives the impression of showing clear advantages of OLSR, although it does not explicitly say so. [36] contains a detailed comparison of OLSR with AODV, which is in favour of OLSR in most (but not all) cases.

7.5.10 STAR vs. topology broadcast, ALP and DSR

STAR is compared with a simple topology broadcast, ALP[10] and DSR in [59], which shows STAR being superiour to the compared protocols (but again, being from the same author as STAR).

---

[6]Directional Routing
[7]Most Forward within Radius
[8]Distributed Bellmann-Ford
[9]Idealized Link State, a purely academic algorithm
[10]Adaptive Link State Protocol

7.5.11   WAR vs. DSR

[3] and [4] compare WAR with DSR (analytically and with experiments) and show under which conditions WAR outperforms DSR. Both papers are from the same author as WAR.

7.5.12   WRP vs. DBF, DUAL and ILS

WRP is compared to DBF[11], DUAL[12] and an ILS[13] protocol in [114], with the results being clearly in favour of WRP.

---

[11]Distributed Bellmann-Ford
[12]The EIGRP routing protocol by Cisco
[13]Idealized Link State

Chapter 8

Description of Individual MANET Routing Protocols

This section briefly introduces the considered routing algorithms for mobile ad hoc networks in alphabetic order. Each section describes one proposed protocol, gives references to specifications, evaluations and studies, which concern that protocol and includes a graph, that shows the degree of similarity of this routing protocol to the other protocols. More details about the similarity function used, can be found in section 7.4.

8.1   ABR - Associativity Based Routing

ABR [156] is an on-demand routing protocol: Routes are discovered with a *Broadcast Query* request. From these requests, the destination learns all possible routes, and replies along a selected route to the source.

If a route breaks, several route-reconstruction methods can be applied, depending if the source, the destination or an intermediate node moves out of reach.

Further, ABR maintains a "degree of associativity" in form of associativity ticks. These are not clearly defined, but from context it appears that every node maintains a tick-value for every one of his neighbors. Every time interval a link-layer hello message from that neighbor is received and the tick value is increased. If the neighbor moves out of reach, the value is reset to zero. A tick level above a certain threshold indicates a stable association between those two nodes.

On selecting a route, the destination (which does the selection) prefers most stable routes, i.e. those with the highest associativity tick value. Hence, this "degree of associativity" is used as a metric of mobility. This strategy is similar to SSA (cf. section 8.24).

In [156], there are statements about the complexity of ABR, but since they lack a clear definition, they may not be very useful.

In [100] ABR was compared to DSR and DBF by a simulation study using GloMoSim in a small scenario[1]. The results are in favor of ABR, in terms of overhead, throughput and end-to-end delay, although the advantage to DSR is quite small. Other criteria, like memory requirements for the table and power consumption show disadvantages of ABR.

ABR is also described in [155], which focuses on the impact of HELLO-messages (beacons) on the battery life of nodes.

1999 ABR was submitted as a draft to the IETF MANET working group under the title "Long-lived Ad-Hoc Routing based on the concept of Associativity". However, the draft has expired since[2], so one can assume the topic was no longer of interest for the working group, and now it even seems that C.K.Toh has abandoned his work on ABR (possibly in favor of more promising methods).

However, in 2001 [112] was published, which proposes an enhancement to ABR. The stability property, again measured in *ticks* is now determined in a more advanced and improved way. Further an optimized threshold for associativity is introduced, such that no longer the route with the highest degree of associativity is chosen, but the one closest to the optimal threshold value. It was claimed, that this OABTR called protocol was compared to DSR (cf. section 8.10) and ABR, but no details about this evaluation have been given.

8.2 ADV - Adaptive Distance Vector Routing

ADV, the Adaptive Distance Vector Routing Algorithm by Boppana and Konduru [23] is described as a combined proactive and on-demand type of protocol. The main characteristic is proactive, since routes are maintained all the time. The on-demand character is implemented by two key aspects:

- only routes to *active receivers* are maintained

- the frequency and size of routing updates is adapted to the current network conditions.

---

[1]30 nodes in a 20 × 20m area.

[2]Six months after submission, a draft must either be updated or be submitted as RFC or Internet standard, otherwise it will expire.

Of course *active receivers* must be announced in a broadcast-like fashion, similar to broadcast route requests. Also, if a node ceases to be a receiver, this must be announced, too. Every node keeps a *receiver flag* for each destination in its routing table, to reflect the status of this node.

To adapt the frequency and contents of routing updates to the network load and mobility, a *trigger meter* is kept by each node. This variable can be increased in certain steps, depending on the events that the nodes receive. There are two thresholds, the first is a dynamic threshold, which is computed on the recent past and the role of the node (e.g. if the node is part of an active route, etc). If this dynamic threshold is exceeded, a partial update is scheduled. The second threshold is a fixed constant TRGMETER_FULL, which will trigger a full update, if it is reached. The trigger meter is reset after each update.

In [23] ADV was compared against DSDV, AODV and DSR, and outperformed any of these in most considered performance metrics (cf. appendix A). In [49] ADV was compared against AODV and DSR, but explicitly with TCP traffic and various TCP modifications to improve TCP performance in mobile ad hoc networks. This paper is in favor of ADV in several terms. ADV is much better if there is significant background traffic. Since it does not benefit much from TCP enhancements like fixed RTO, SACK and delayed

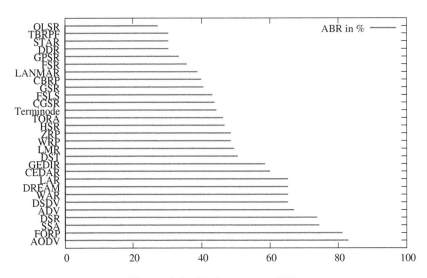

Figure 8.1: Similarities to ABR

ACK (but performs equally well as AODV or DSR with these enhancements), it is argued, by using ADV such special TCP features may not be needed.

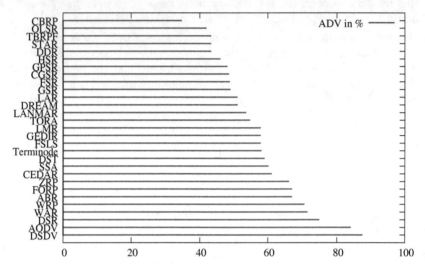

Figure 8.2: Similarities to ADV

8.3   AODV - Ad Hoc On Demand Distance Vector Routing Protocol

This is one of the most discussed and most advanced routing protocols. It is an important part of the work of the *MANET* IETF working group. The draft was recently accepted as experimental RFC 3561[125]. So this is probably the most mature suggestion for an ad hoc routing protocol. Its main developers are Charles E. Perkins (Nokia Research) Elizabeth Belding-Royer (UCSB) and Samir Das (University of Cincinnati). AODV is discussed in lots of studies and is often used as a reference to compare other routing protocols. AODV was derived from C. Perkins earlier work, DSDV (cf. section 8.9). Compared to DSDV, AODV no longer needs to exchange periodic messages proactively, but works in an on-demand fashion, instead.

If a route to a destination is unknown, a route discovery process is initiated. This consists of broadcasting a Route Request (RREQ) packet throughout the network. To limit the impact of a net-wide broadcast, these request should be sent with an expanding ring search technique: the TTL of the packets starts with a small value; if no route has been found, the TTL will be increased and the request will be resent. Each node that rebroadcasts this request, adds its address into a list in the packet. If the destination sees the request, it will reply with a unicast Route Reply (RREP) to the source. Each intermediate node may cache the learned routes.

The routing table entries consist of a destination, the next hop toward this destination and a sequence number. Routes are only updated if the sequence number of the updating message is larger than the existing one. Thus routing loops and updates with stale information are prevented. The sequence number technique was already used in DSDV (cf. section 8.9) and was adopted by a variety of other routing protocol developers.

The amount of information, which needs to be present at each node, is rather limited:

- The node is aware of its neighbors (via link-layer-notification, or explicit HELLO messages).

- The node knows route destinations and the next hop.

- The node has a "precursor list" for each destination. This list consists of all nodes, which use the current node as a relay for the destination. In case of a route failure to this destination, the node knows exactly which other nodes to notify.

- Each routing entry also has a lifetime.

The authoritative description of AODV is RFC 3561[125].

A more easy to read description is given in [128]. A huge number of ad hoc related papers cite AODV as a reference (we do not list them here). However, some papers did an independent comparison between some ad hoc routing protocols including AODV, like [25], [85] and [45]. In these papers AODV and DSR compete better than other protocols and AODV shows the best results overall.

8.3.1 MAODV

There is a special form of AODV for multicast traffic, called MAODV. I did not have a close look at MAODV and just mention the main differences in a rough way:

- Routing tables have more than one next-hop.

- Route discovery is initiated on joining of a group, or on sending a message to a group with no route.

- On join requests, only a multicast router or a tree member should respond, otherwise, any node with a route will do.

8.4 CBRP - Cluster Based Routing Protocol

CBRP maintains clusters of two hops diameter, with an elected clusterhead for each cluster. Clusters may be overlapping, but each node must be part of at least one cluster. Clusterheads are not allowed to be direct neighbors, except for a short period (called "contention period"). Nodes maintain a neighbor table which also includes the link type. Also a cluster adjacency table is kept in each node. Source routing is used, with the route in the CBRP header. This allows a limited local repair mechanism and a route cache (much like DSR, see section 8.10) to be used.

For clustered routing, the key argument is that with a clustered hierarchy, it is again possible to channel information (cf. Section 7.2.3). Thus scalability may be regained, even if broadcasts need to be used.

This routing protocol was submitted as a draft to the IETF MANET working group in 1999 [84]. This draft is now also expired, but CBRP is also described in [148].

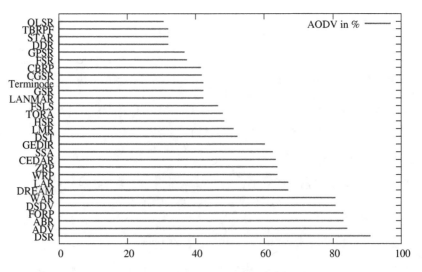

Figure 8.3: Similarities to AODV

Unfortunately there does not seem to be much more work published about it.

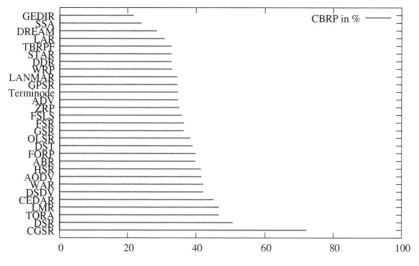

Figure 8.4: Similarities to CBRP

8.5 CGSR - Clusterhead Gateway Switch Routing

In [35], Clusterhead Gateway Switch Routing is proposed. It consists of a clustering method, called *Least Cluster Change* which is combined with either "lowest id", or "maximum links", to form clusters and elect clusterheads. The method focuses on cluster stability. CGSR explicitly specifies requirements on the link layer and medium access method:

- Inter-cluster communication requires a CDMA system, such that each cluster is assigned a different code (spatial reuse of codes is utilized, though).

- Within each cluster, TDMA is used. The allocation of time slots is done by a token passing method.

Gateway nodes are nodes, that are within more than one cluster, and therefore need to communicate in different codes.

The protocol uses a sequence number method (as developed in DSDV) to gain loop-free routes and avoid stale routing entries. In CGSR, a packet is routed alternating between clusterheads and gateways, hence the name. In the paper, several enhancements (e.g. priority token passing) are suggested, as well.

Simulation of the protocol was done by using a special simulation language called *Maisie*. A 500 × 500m region was used, with 100 nodes. The nodes did move according to a random strategy that was no further specified.

CGSR is mentioned in [136] and [57]. Implementations for common simulators, or even real-world use, don't seem to exist.

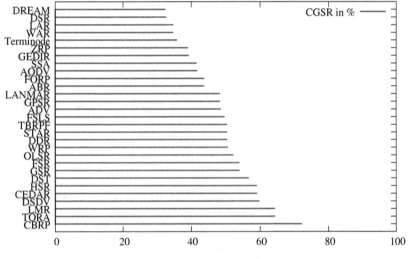

Figure 8.5: Similarities to CGSR

8.6 CEDAR - Core-Extraction Distributed Ad Hoc Routing

This is more a routing framework method for QoS requirements than a MANET routing protocol. In CEDAR a subset of the nodes is selected that will form a backbone within the network (the *core*). This structure is used for broadcast messages, hence no flooding is needed. The messages sent over the core network are *increase waves* (slow propagating) and *decrease waves* (fast), which notify about an increase or decrease of available bandwidth.

The propagation of these waves is dynamically limited, depending on the available bandwidth. So the relevant information for QoS is disseminated in an efficient way. Within the core network, any established ad-hoc routing protocol may be used. The usage of this information, in order to establish QoS routes, works as follows:

A node contacts its "dominator" (local core node) with a route request, that contains source, destination and required bandwidth. The dominator computes a QoS route, if this is feasible and then continues to establish it. This includes possible discovery of the dominator of the destination and a core path to it.

CEDAR was presented in [144]. Its QoS focus, and also the proposal of a core/backbone network distinguishes CEDAR from most other routing strategies, and it is often mentioned and cited in other papers.

[143] picks up the idea and suggest a way how to improve performance of AODV or DSR by the use of a core infrastructure as proposed in [144].

However, it does not seem to have caught enough interest such that subsequent practical work, like implementations of CEDAR for simulators or real environments and evaluations, would have been done.

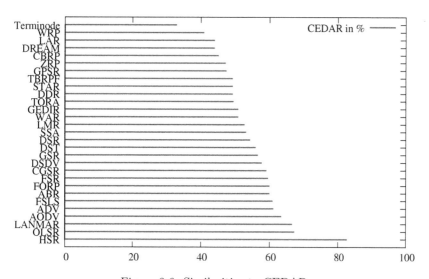

Figure 8.6: Similarities to CEDAR

8.7 DDR - Distributed Dynamic Routing Algorithm

DDR is based on the construction of a forest to represent topology, which is constructed by using local periodic messages only. (This is a similar approach as in OLSR, see section 8.23.)

DDR also forms a set of disjoint routing "zones" (cf. section 8.31). There is a zone for each tree in the forest. Routing information is exchanged only with nodes, that are within a node's zone and which concerns only neighbor zones. The zone size is not fixed in DDR but will be adjusted dynamically. The algorithm in detail is given in [115], but no simulation or any other performance comparison was done so far.

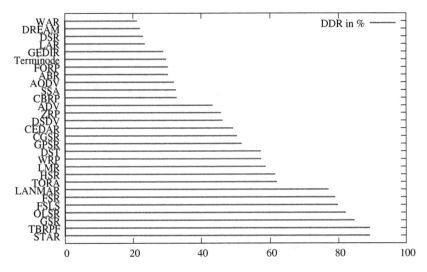

Figure 8.7: Similarities to DDR

8.8 DREAM - Distance Routing Effect Algorithm for Mobility

This algorithm was suggested in [10] at MobiCom 1998. It is a location based algorithm, that makes use of the *distance effect*. This means, that two nodes appear to move slower with respect to each other with increasing distance. Thus, location information for distant nodes does not need to be updated in such an accurate and frequent way, as for close nodes (see also FSR in section 8.14 for a similar approach).

Each node has a routing table with location information about each other node. DREAM can be considered proactive, since location information must be disseminated (the method of location determination is not specified, so a separate location service[3] may be required).

On sending a message, a *direction* is determined by using the location of the destination. Then, the message is passed to all neighbor nodes in that direction. This method is more related to reactive protocols, as the route is not fixed in advance.

Distance and mobility of a node determine the frequency of location updates. A fast moving node sends location control messages much more often than a slow one. Also the messages are sent with a different maximum amount of hops (also called time to live) and short-lived messages are sent much more frequent than long-lived. The long-lived messages will reach far away nodes, but are sent much less frequent. This leads to a bandwidth and energy efficient protocol. Although the routes are not fixed in advance, there is no setup-delay.

Basagni et. al. claim that this protocol is inherently loop-free, since the messages travel away from the node into a specific direction. This could be questioned, since in a network with very high mobility, the target direction can change, even back to a node who has sent the message already. There is also some discussion about this in the papers about GEDIR, cf. section 8.15.

Another problem is, that location table entries may be stale and that no close neighbor in the required direction can be found (e.g. due to lack of connectivity). Both problems are addressed in [10], but not very detailed: an unspecified `Recovery()` routine should be called in such cases. The authors chose to use flooding in their prototype implementation.

There was not much more work on DREAM, but other routing methods such as LAR (cf. 8.20) or FSR (cf. 8.14) did pick up some concepts of DREAM.

---

[3]GPS for instance can serve only as a component in such a location service, since it can only provide location data for each individual node. The nodes must still announce or distribute their location.

8.9 DSDV - Destination Sequenced Distance Vector Routing Protocol

This protocol is the result to adapt an existing distance vector routing algorithm (Distributed Bellman Ford, [14]), as used in RIP, to an ad hoc networking environment. This is a proactive protocol, that updates routing information on a regular basis. To avoid routing loops, destination sequence numbers have been introduced. DSDV is one of the first attempts to adapt an established routing mechanism to work with mobile ad hoc networks.

Each routing table lists all destinations with their current hop count and a sequence number. Routing information is broadcast or multicast. Each node transmits its routing table to its neighbors. Routes with more recent sequence numbers obsolete older routes. This mechanism provides loop freedom and prevents stale routes.

The routing information is transmitted every time a change in the topology has been detected (i.e. a change in the set of neighbors of a node).

DSDV works only with bidirectional links.

DSDV was presented in [126] in 1994. A more detailed description is available in [127]. DSDV was also used for many comparisons like [25], [44], [85] and

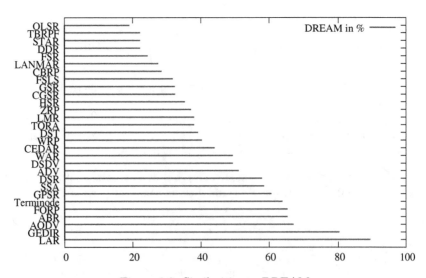

Figure 8.8: Similarities to DREAM

[23]. The results are mixed, but especially the later papers show results, where DSDV is not performing well compared to the other protocols including AODV which was developed by the same author.

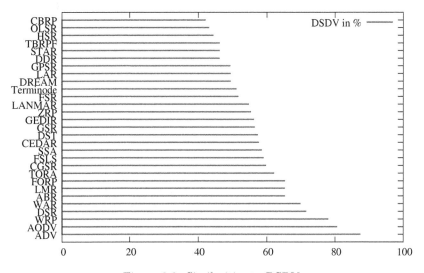

Figure 8.9: Similarities to DSDV

8.10   DSR - Dynamic Source Routing

DSR is an on-demand protocol, that uses source routing. In this case, this means, that each packet carries the complete route to its destination in its header (which introduces some overhead). It was first described in [86]. Since DSR works on demand, a route must be discovered through a *Route Discovery Mechanism* before use. Discovered routes may be cached and routes may be *overheard* by a node (by parsing the source route information of packets that are relayed).

If broken links are detected, a corresponding *Route Error* message is transmitted through the network and a route maintenance mechanism takes over to fix the broken routes, if possible.

To further reduce unnecessary traffic, a node may reply to a route request with a locally cached route, even if it is not the destination node. Delays in these replys with promiscuous observation (*overhearing*) of other routing traffic prevent multiple nodes replying with a cached entry all at once.

The dynamic source routing protocol is also a very mature protocol. The IETF draft [88] has reached version 7 and will result very likely in some IETF standard (probably experimental RFC) as well as AODV.

DSR is described in detail in [87]. DSR is also one of the few ad hoc routing protocols, that have been implemented and evaluated in a real testbed. The results are described in [111].

[4] presents an analytical study of the probabilities of successful deliveries and the total amount of traffic generated for a successful delivery. It is argued, that an end-to-end recovery mechanism (as used in DSR) does not scale if the routing path lengths increase. Instead a local recovery mechanism is suggested, that gives much better results (according to the analysis). This algorithm is implemented in the Witness Aided Routing Protocol (WAR, cf. section 8.29). Although the analysis is convincing, it was done by the authors of WAR, so the result being strongly in favor of WAR is not surprising.

DSR was used in many performance comparisons, evaluating studies, and was used as a reference for a lot of other protocols. Further, it was used as a reference protocol for investigations to find general improvements for mobile ad hoc networks (like reduced energy consumption). Papers referring to DSR include: [25, 110, 44, 53, 136, 85, 43, 45, 170, 49, 4].

[100] also compared ABR to DSR and DBF. The result is that both ABR and DSR perform much better than DBF, with a slight advantage of ABR.

In [147], a closely related protocol called *Neighborhood Aware Source Routing, NSR* is described, which is based on the DSR ideas. NSR is not considered further in this study.

8.11   DST - Distributed Spanning Tree Protocol

This approach takes into consideration that in a mobile ad hoc environment, there can be regions of different stability. So this approach proposes the establishment of a backbone network in the stable regions, using a spanning tree algorithm.

For the unstable regions a flooding or a so called *shuttling* approach is used to transmit the packet to the destination even through a very unstable area.

DST is described in [131] and compared against pure flooding.

There was no comparison to other approaches and this paper is also only mentioned briefly in [115].

8.12   FORP - Flow Oriented Routing Protocol

FORP[150] is designed for real-time traffic flows (over IPv6). It works in an on-demand fashion (similar to other corresponding protocols), such that

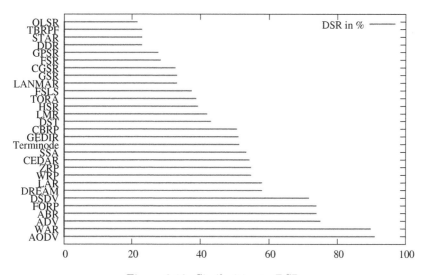

Figure 8.10: Similarities to DSR

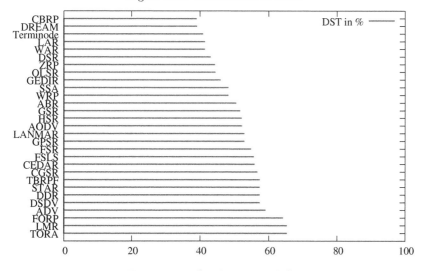

Figure 8.11: Similarities to DST

traffic flows[4] are at first requested and can then be used, if granted. The characteristic of FORP is that for each link there is a Link Expiry Time (LET), and the minium of all LETs for all links in a route gives the Route Expiry Time (RET). [150] suggest, how these expiry times can be predicted and [151] discusses route prediction in more detail. Just before a link or route expires (i.e. a critical time is reached), the destination sends a Flow-HANDOFF message, which triggers another Flow-REQUEST, thus finding a new route over which the current flow can be rerouted, without interrupting it. There have been no comparisons or any more work about FORP to my knowledge.

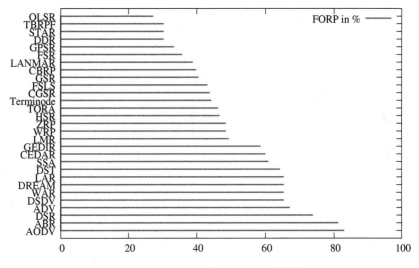

Figure 8.12: Similarities to FORP

8.13   FSLS - Fuzzy Sighted Link State Algorithms

This class of algorithms, as described in [138] also addresses the problem of the limited dissemination of link state information, similar to DREAM or FSR (cf. sections 8.8 and 8.14). LSUs (link state updates) are sent with a dynamically limited maximum hop count (time-to-live), and in certain intervals, which depend on the number of hops, the updates can travel. Far

---

[4]more precisely the routes for a flow

reaching LSUs are sent much less frequent than short reaching LSUs. Also LSUs are only created if the state of a link has changed within the scope of the LSU. The length of the intervals and scope of the LSUs are the design parameters of the class of FSLS algorithms. An extreme case is the *discrete link state* algorithm DLS, in which each LSU is sent through the whole network (TTL is set to ∞). It differs from standard link state only in the fact, that the LSU is not sent immediately after a link status changes, but at the beginning of the next interval.

[138] also derives an optimal case for a FSLS algorithm, the *Hazy Sighted Link State Algorithm* (HSLS). [139] goes into more detail about both.

In [139], a more comprehensive overhead definition, which includes overhead due to non-optimal routes, is used to analyse the class of FSLS, derive the HSLS and prove its optimality.

However, for the analysis some assumptions are made that may not be the case in certain scenarios (e.g. the traffic a single node generates is independent of the network size). They are meant to apply for the average scenario, without considering border effects.

More analytical studies about FSLS and overhead due to non-optimal routes can be found in [137] and [80].

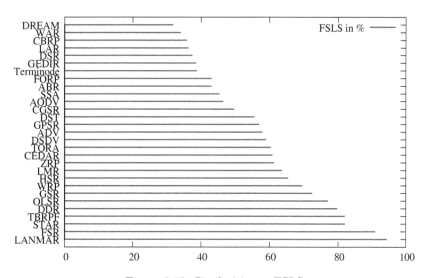

Figure 8.13: Similarities to FSLS

8.14 FSR - Fisheye State Routing

Fisheye State Routing was proposed by Mario Gerla et.al. to the MANET IETF working group (see draft [63]).

Similar to DREAM [10] FSR wants to reduce unnecessary traffic by introducing a multi-level scope. By concept, FSR is a protocol that periodically updates link state information (table driven).

FSR is derived from Global State Routing (cf. section 8.17). The major drawback of GSR is the large message size and the propagation latency of the link state changes. FSR now helps by introducing *scopes*, which depend on the number of hops a packet has reached from its source. Nodes within the smallest scope are considered most often in update packets; nodes, which are far away are considered much less frequent. This means, the message size can be greatly reduced, as information for most nodes can be omitted. Although routes may become inaccurate for distant destinations under increased mobility, packets will find more and more accurate routes while getting closer to the target, thus they don't suffer much from the inaccuracy. FSR is explained as well in [78], while [153] reports about the implementation of FSR.

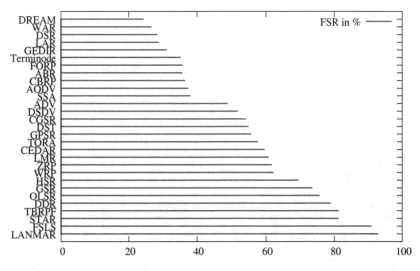

Figure 8.14: Similarities to FSR

8.15 GEDIR - Geographic Distance Routing

This approach uses geographical information, like DREAM and LAR (cf. sections 8.8 and 8.20). While these protocols use directional routing (i.e. a message is sent to one or more neighbors *in the direction* of the target), GEDIR uses an approach based on progress (in terms of the proximity of the packet to its destination) to select the set of neighbors to forward the message to. [104] describes a set of related geographic routing protocols (Directional Routing - DIR, Most Forward Within Radius - MFR and GEDIR in some variations) and their advantages and disadvantages.

The authors of [104] show by a counterexample, that loop-freedom for DREAM does not hold and they intend to show, that their methods are loop free, provided loops are not formed intentional. The situation described in the counterexample appears very artificial and may not appear in practice at all. Also no statement is made about the duration of loops. However, in a static network without any movement the formation of such loops is much more likely than in a dynamic network.

The result of [104] is, that most of them perform well under certain conditions. Their suggested algorithm (GEDIR) performs best among the discussed ones. Another interesting result is that multiple paths, as provided by many geographic routing protocols, do not improve the overall delivery ratio very much.

Alas, [164], an earlier paper about GEDIR, is no longer available.

8.16 GPSR - Greedy Perimeter Stateless Routing

GPSR [89] is another location based routing protocol. A node learns about the position of its neighbors by a beacon or by information piggy-backed on data packets (similar to other neighbor discovery methods).

The node forwards a message in a *greedy* way, i.e. to the neighbor which is geographically closest to the destination. If there is no such neighbor (which means, the node itself is currently the closest node to the target within its transmission range) and the target is not in range, GPRS switches to *perimeter mode*, which guides the packet around this void area, using a planar-graph traversal with the right-hand rule (Chapter 3 in [89]). On entering perimeter mode, the current location is registered in the packet, such that greedy forwarding can be resumed, as soon as the void is traversed. Of course GPSR requires that each node is aware of its own position, possibly by means of a GPS device. Also it is required that any source node knows the location of its destination. This information is registered once in the packet and never changed. Karp and Kung do not really address the problem of how to obtain the target location for a source node in [89]. There is only a reference to a location database service, which needs to be looked up. Hence GPSR depends on such a service like GLS[101], which may not be available in a mobile ad hoc network. However, GPSR was not designed just to work in a mobile ad hoc network, but also in rooftop and sensor networks. GPSR is certainly a promising approach, but like other position based protocols, does not solve the problem of a location service, which must be available, thus limiting its applicability.

Performance of GPSR was compared against DSR using the NS-2 network simulator and showed better results than DSR for the environment used in the evaluation.

GPRS is described in detail in B. Karp's PhD-Thesis [90] and mentioned in a large variety of papers, as a reference. Few papers, like [24] and [91] go more into depth, but there is none, which does a detailed comparison with other protocols[5]. [102] and [57] show (as a corollary) that perimeter mode can be improved if a *Delaunay Graph* is used instead of a *Gabriel Graph.* [9]

---

[5]Except [168], but that focuses on energy conservation and sensor networks.

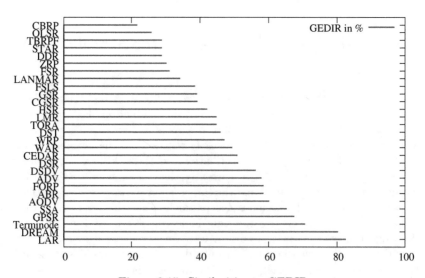

Figure 8.15: Similarities to GEDIR

presents an improved routing algorithm based on GPSR.

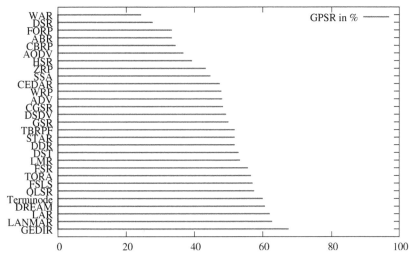

Figure 8.16: Similarities to GPSR

8.17   GSR - Global State Routing

Global State Routing[32] was developed by Tsu-wei Chen and Mario Gerla. It is an early attempt to introduce link state routing in an ad hoc networking context. The main problem of traditional link state routing is the high amount of topology information, which is sent from each router to each other router. Since in ad hoc networks, each node is also a router, this mechanism does not scale and needs to be optimized. Therefore GSR adopts the information dissemination process from the Distributed Bellman-Ford algorithm[14]. In GSR topology information is exchanged periodically only among neighbors. If a topology change occurs, this change is transmitted further. Messages are sent only in such triggered cases. GSR uses sequence numbers based on timestamps, but no method of clock synchronization of the nodes is suggested or even mentioned[6].

GSR was evaluated in simulation and compared to traditional link state and Distributed Bellman-Ford. There are no other comparisons or any further work with GSR, that I am aware of. Since the research group of Mario Gerla

---

[6]Synchronous clocks are probably silently assumed.

has developed a wide range of other routing strategies since the development of GSR, it can be assumed that further work on GSR has been abandoned in favor of more superior approaches, like FSR or HSR (cf. sections 8.14 and 8.18).

Although GSR is cited in some papers, these papers mention it only as an example.

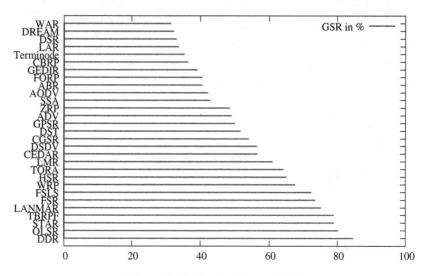

Figure 8.17: Similarities to GSR

8.18 HSR - Hierarchical State Routing

HSR[78] introduces a multilevel clustering infrastructure. Clustering is done on a physical and logical basis.

Physical clustering starts with level 0, the bottom layer. On each level, nodes can form clusters, which are represented by a clusterhead node each. The clusterheads itself can form another cluster at the next higher level. On higher levels the clusterheads are connected via *virtual links*, which need to be mapped to physical links on the bottom layer. A virtual link will usually contain gateway nodes on the lowest level.

Each clusterhead collects link state information of each cluster member and about their neighbors and propagates a summary to its fellow clusterheads on the higher levels, possibly using gateway nodes. On the higher levels the same happens with the link state information about the virtual links, which are computed from the lower level link states.

A special hierarchical addressing method is used, which is sufficient to route a message from any node to any other node[78]. A node passes a message up to the node of the highest level in its current hierarchy. This one will pass it to the destination cluster node (through a virtual link), which will pass it down the levels to the right node on the lowest level.

Additionally to physical clustering, a logical partitioning is used, which works similar to Mobile IP. The details can be looked up in [78].

HSR (as well as FSR, cf. section 8.14) claims to be "QoS ready" [78], since QoS criteria can simply be taken into account into the link state and both protocols operate as link state protocols.

In the evaluation section in [78], a channel allocation method was used such that the cluster heads poll each cluster member and subsequently assign channels to them on demand. This can even be used with the WLAN standard IEEE 802.11 using the PCF instead of the DCF.

The evaluation compares HSR, FSR, DSDV and two on-demand routing methods, which are not specified in detail. The results are not as explicit, as one could have expected. Still HSR shows its advantage as being the most scalable approach, FSR instead (which is also proposed in this papers) does not perform equally well.

The paper is cited by various other works, including [152], an analytical study about clustering overhead in general, [138] (cf. section 8.13), [61] (cf. section 8.19). In [65] an enhancement to HSR, now called EHSR is proposed, designed for the military.

8.19 LANMAR - Landmark Routing Protocol

LANMAR [123] is the result of combining FSR (cf. section 8.14) with Landmark routing [157]. LANMAR combines both link state and distance vector characteristics. LANMAR utilizes the landmark routing with group mobility, i.e. groups of nodes, that are likely to move together build a subnet. In each subnet a landmark node is elected. Compared to FSR, only the information about nodes in scope (within a subnet/group) and those of the landmarks are transmitted in the link state updates. If a packet needs to get routed to a distant node, it is routed to its landmark node. As soon as it gets within scope of the destination node, it will get a more accurate route to the destination. It may not even be required to route the packet through the landmark.

The link state update process is very similar to FSR with the addition of a distance vector which is determined by the number of landmarks (logical subnets). Also there is now a fixed scope (all nodes within the scope are fixed and not determined by distance), and the update interval is now constant. Paths are kept to nodes within a subnet that travel out of scope of their landmark. The extra overhead to keep track of these drifters is shown to be relatively low. Another problem can be the existence of isolated nodes, which belong to no group, but could be their own landmark. Depending on the fraction of such isolated nodes, special handling may be required (e.g. reverting to traditional FSR).

LANMAR was evaluated per simulation and compared against plain FSR, as well as DSR and AODV. The results are in favor of LANMAR, especially in the cases with many nodes and high mobility. Also LANMAR clearly outperforms FSR in these simulations. It has to be made clear, that the scenarios given do explicitly use a group mobility model of which LANMAR can get a high benefit.

In [61] LANMAR is extended by a landmark election process, which was not specified in [123].

[123] is cited in a few papers including [99], but no further analysis or eval-

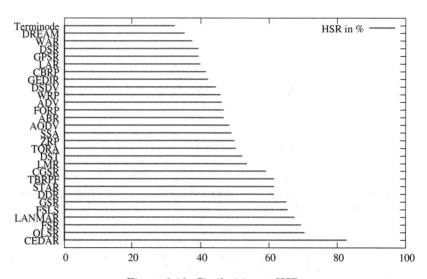

Figure 8.18: Similarities to HSR

uation was done (to my knowledge).

LANMAR was also submitted as an Internet Draft[62] to the MANET IETF working group.

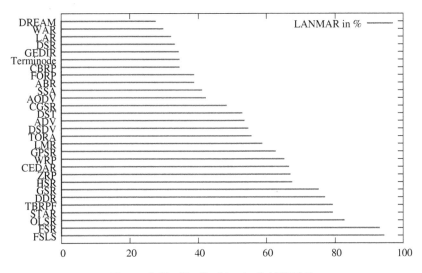

Figure 8.19: Similarities to LANMAR

8.20 LAR - Location Aided Routing

Location Aided Routing, as proposed by Ko and Vaidya [93], is an enhancement to flooding algorithms to reduce flooding overhead. Most on-demand methods, including DSR and AODV (cf. sections 8.10 and 8.3) use flooding to obtain a route to the destination. This flooding results in significant overhead. LAR now aims to reduce the overhead to send the route requests only into a specific area, which is likely to contain the destination.

For this purpose the notions of *expected zone* and *request zone* is introduced. The *expected zone* covers the area, where the destination node is expected, according to the currently known information like:

- location at some time t (this will be the center of the expected zone)

- speed at time t

- direction at time t

Of course, this extrapolation of the state of the node at time t, does not need to be accurate at some later time t', but it provides a good start. Since the expected zone does not need to contain the source node, a larger area than the expected zone must be covered by the flooding, including all possible nodes on the way from the source to the expected zone. This expanded expected zone is called *request zone* and is used to restrict the flooding, i.e. only nodes that are part of the *request zone* forward a route request. On unsuccessful route discoveries, the request zone may need to be expanded further, possibly covering the whole network. Such subsequent route requests increase the initial latency for connections. This results in a tradeoff between reduced overhead and increased latency, and needs to be balanced carefully. Depending on the method used, a sender needs to include a specification of its request zone in its route request such that nodes receiving the request, can determine, whether they are within the zone or not. A node replying with a route will include its coordinates along with the current time (and possibly other parameters like speed and direction) in the reply, so that the sender will have its coordinates (at that time) for future requests.

LAR was evaluated with MaRS [160] in a couple of scenarios, but the authors of [93] just compare two different modes of LAR, but no other routing protocols.

Suggested improvements include adaption of the request zone on the fly by the intermediate nodes of the route request. More flexible forms of request zones may be used and location information can be piggy-backed to any node, to keep location information more accurate within the network.

[93] is one of the most cited papers in research area of mobile ad hoc networks, but it is commonly used only for reference by related work of other authors. [94] is a subsequent paper, that shortly emphasizes on a few optimizations of LAR.

8.21 LMR - Lightweight Mobile Routing

Lightweight Mobile Routing is a link reversal routing (LRR, cf. section 8.22) algorithm, that was developed to overcome the non-convergence problem in partitioned networks with the previous methods as proposed by [55]. LMR was published in [39] and also described in [40].

The scope is generally a scenario, where changes happen too frequently for link state algorithms to adapt to, but not that frequently, that flooding is the only choice. LMR focuses on low complexity instead of optimal paths, such that it can even scale in very large networks.

Like many other protocols, LMR also uses three basic messages: QRY (query),

RPY (reply) and FQ (failure query). They correspond to the messages used in AODV, DSR (cf. sections 8.3 and 8.10) and many others.

A QRY is sent by the source node using a limited broadcast (see appendix A). The source then waits for a RPY packet, which will be issued by any node, which has a route to the destination and received a QRY or FQ packet. The *directed flood* caused by the RPY messages forms a directed acyclic graph (DAG), rooted in the originator of the RPY. The route itself and the up- and downstream links formed depend on the order of the RPY transmissions.

If a node loses its last route to the destination and it has upstream neighbors (cf. *precursors* in AODV), a FQ is broadcasted, to erase invalid routes. On reception of a FQ, the node may either transmit a RPY (if it still has another route) or another FQ if its last link was erased by the first FQ. So instead of a direct link reversal, LMR erases the links and sets them up new.

Loop freedom in a dynamic environment is ensured by marking previous unassigned links as "downstream-blocked" if the node has already an up-stream link. These markers time out after a while, but it may happen that a downstream link cannot be used, because of possible loop formation. A similar mechanism is used to avoid deadlocks.

LMR is mentioned in a large number of papers, but only as a reference. LMR became less interesting with the development of TORA (cf. section 8.28) as

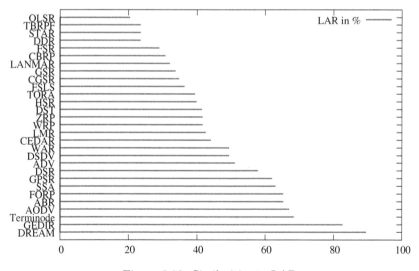

Figure 8.20: Similarities to LAR

a successor.

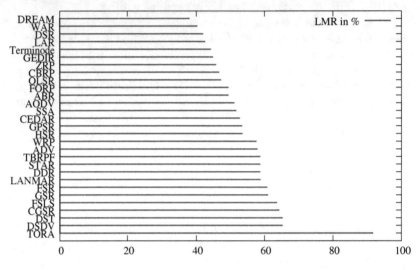

Figure 8.21: Similarities to LMR

8.22 LRR - Link Reversal Routing

As described in [40] LRR is a certain routing approach for highly dynamic networks. Its objective is to minimize the amount of overhead, when topology changes need to be announced. The maintained topology is reduced to a directed acyclic graph (DAG), rooted in the destination.

As the graph is directed, each link is either upstream or downstream to the destination. If a node in the graph becomes a local minimum, i.e. it has no downstream, one of its links is reversed. To achieve this, a notion of *height* is introduced, thus the problem is similar to flows in a graph. The height of the minimum node is raised such that it is higher than the lowest of its neighbors, thus reversing the direction of this link. The reversal can cause another node to become a minimum and the process continues.

The drawback is that no node knows about the "distance" (in any term) of itself to the destination, so optimizing metrics, as used in distance vector or link state algorithms, cannot be used.

LRR itself may be used in a proactive or reactive way.

The first and simple approach for LRR is *Gafni-Bertsekas'* Algorithm [55]. Further development lead to LMR and TORA (cf. sections 8.21 and 8.28).

8.23 OLSR - Optimized Link State Routing

OLSR is another proactive link state protocol, which is claimed to work best in large dense networks.

Each node selects a set of "Multi-point Relays" (MPRs) from its neighbors. The radio range of the MPR set should cover all 2-hop neighbors. Each node knows for which node it acts as a MPR. Thus OLSR requires bidirectional links. OLSR distributes routing packets via UDP. Each routing packet contains one or more OLSR messages. Messages exist for neighbor sensing, topology declaration and MPR information, interface, host- and network declaration.

OLSR explicitly requires avoidance of synchronous packet emissions among nodes in the neighborhood, to reduce channel competition, which is probably a unique explicit requirement for an ad hoc routing protocol. For this purpose, jitter is used during transmission periods.

From the topology information, a shortest path for each destination is computed.

OLSR was first introduced as an IETF draft to the MANET working group in 1998. The draft has evolved since and was accepted as experimental RFC 3626[38]. The first draft was cited in several papers, but none of them goes into much detail. There were few performance comparisons: [79] which does an analytic comparison of OLSR with DSR in a random graph model, and [36] did a very detailed comparison of OLSR with AODV, which is mainly in favor of OLSR (but not in all cases).

OLSR has some similarities to TBRPF (cf. section 8.26)

8.24 SSA - Signal Stability-Based Adaptive Routing

SSA [48] presents a totally different approach from most other routing algorithms. The focus is to use signal and location stability as main routing criteria. The routing framework behind that works like most on-demand routing algorithms, i.e. route requests are broadcast through the network, route replies are returned by the destinations, routes are set up accordingly. The stability criteria interact with the standard procedure like this:

Each packet received is first passed to a module called DRP (dynamic routing protocol). DRP interacts with the device driver of the network interface using an API, that allows to pass signal strength information. DRP maintains a *signal stability table* and categorizes each link to its neighbors as either being *strong* or *weak*. This table is updated with every packet received. Beacons

(HELLO packets) are not processed further, but routing and data packets are passed up to the SRP (static routing protocol) module, which performs the usual routing tasks, like reacting to route request, forwarding packets according to the routing table, etc.

A route request can state, whether it wants any kind of links or just strong links. If only strong links are requested, any node receiving a route request over a weak link will drop it. Thus only route requests over strong links will reach the destination. The destination selects the first route request received by the same originator as the route and sends its reply vie the reversed hop list in the received request. The strategy suggested in [48] is first to try only strong links and fall back to any link, if no route could be found.

Also two enhancements are suggested: an additional link requirement, which just *prefers* strong links over weak ones (but does not rule them out). In this case route requests are all forwarded, but each intermediate node adds the link quality into the route request packet. Also the destination does not choose the first route request it receives, but waits a while to choose the best route in terms of strong links from all the route requests it has received so far for that source. The second improvement is a gracious route reply by intermediate nodes that already know a route to the destination (as in various other proposals, like DSR).

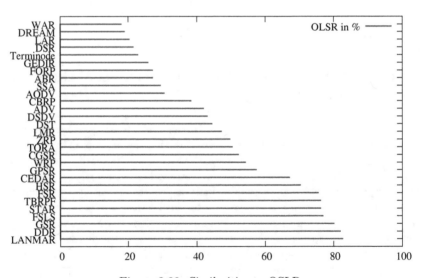

Figure 8.22: Similarities to OSLR

The simulations have been done without stating the simulation software used. Comparison was against a so called "simple routing protocol", which always chooses the shortest path. It is unclear if this should be regarded as an optimal routing algorithm. The result shows some advantages (fewer route repairs need to be done), but also drawbacks (longer routes on average, since not all links can be used, and a short distance between hops is encouraged due to the stability criteria). Overall performance measures like routing overhead, throughput or packet latency have not been considered. So it is very unclear, if there is any benefit at all, or if the advantage of fewer repairs and reduced broadcast is consumed by the longer path-length or multiple route requests.

Signal Stability-Adaptive Routing seems to be related to the concept of ABR (cf. section 8.1), with just some minor differences.

SSA was mentioned in some other papers, but only as an example for this specific routing approach. There were no detailed comparisons or analysis of SSA performed, so far.

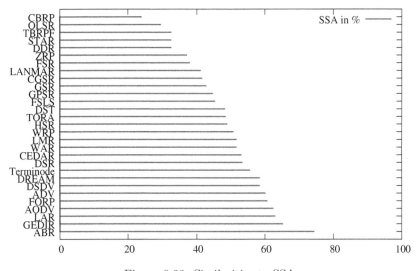

Figure 8.23: Similarities to SSA

8.25   STAR - Source Tree Adaptive Routing

STAR is proposed as an efficient link-state protocol by J.J. Garcia-Luna-Aceves[59]. Each node maintains a source-tree, which consists of its preferred links to each other destination. The source tree is computed on the information of its own links and the source trees reported by its neighbors. Changes in its own source tree are consequently reported to the neighbors. This can be done in an incremental way. The source tree and neighbor information forms the partial topology information in each node. Based on this information a route selection algorithm is run to obtain the route table with destination and next hop.

Information is updated with link state updates (LSU). An update message can contain one or more LSUs, which reflect the changes in the nodes source-tree. Sequence numbers are used to distinguish current from outdated information. The link state information does not time out, thus removing the need for a periodic update.

STAR can operate in several ways. Suggested are two modes: Optimum Routing Approach (ORA) and the Least Overhead Routing Approach (LORA). In ORA shortest path routing is the goal, while in LORA path optimality is not as important as reduced overhead. However the *total overhead*, which includes overhead due to non-optimal paths, as described in [80], [137] and [139] (cf. section 8.13) is not taken into account.

It is claimed in [59] that STAR is the first table-driven protocol, that can use the LORA approach. Other such protocols would need periodic updates to prevent routing loops. This can be avoided by STAR with the use of the routing trees, which can tell any router if a loop may be formed.

STAR requires a *neighbor protocol*, which ensures that new neighbors and leaving neighbors are detected in finite time. Further, STAR requires a link layer, capable of transmitting local broadcast messages without hidden terminal interference. This requirement is not entirely clear, but it seems related to the problems, that occurred with TORA over IMEP in several simulation studies (cf. section 8.28).

However, STAR can still work without this prerequisite, but it is advised to include the whole source tree in each LSU. The broadcasts should then be done in an unreliable (but much more lightweight) way.

STAR was compared against a traditional link state algorithm based on topology broadcast like OSPF, a method called *Adaptive Link State Protocol* (ALP)[58] [7] and DSR (cf. section 8.10).

All simulations are in favor of STAR but I note, that they have been performed by the authors of STAR.

STAR is described also in [60], but the article is very similar to the original in [59].

---

[7]ALP is explained in [60] as *Account, Login, Password*. This seems to be an error and can confuse the reader.

[146] describes ALP, STAR and NSR (cf. last paragraph in section 8.10) in detail including comparisons.

Some further development of STAR lead to SOAR[135][8].

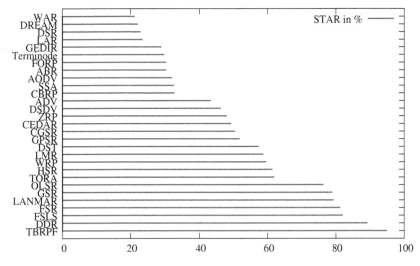

Figure 8.24: Similarities to STAR

8.26 TBRPF - Topology Broadcast Based on Reverse Path Forwarding

TBRPF is a proactive link state protocol, first presented in [12]. It is based on the *Extended Reverse Path Forwarding* Algorithm [42], but does overcome the reliability problems with ERPF.

TBRPF maintains a spanning tree in each node for each other node as the source. This tree is formed by each parent of the *source* node. A list of parents is kept at each node for every other node, as well as a full topology table, including cost and sequence number for each link, the node is aware of. The topology update messages are sent along these spanning trees but in the reverse direction. Of course these updates also will result in modifications of the current spanning tree.

[12] describes only the *full topology* mode of TBRPF. It also provides a proof of correctness (under certain constraints), some complexity analysis and a

---

[8]SOAR is not described in this study.

simulation based performance evaluation.

TBRPF was submitted as an IETF draft to the MANET working group, which has reached version 11 now[117]. Since the original draft, there have been significant changes: A *partial topology* mode was introduced, and in the most recent draft, this is also the default operation (*full topology* mode still exists as an option).

TBRPF supports only bidirectional links. The topology updates are transmitted reliable (i.e. they are acknowledged). A HELLO message is used for neighbor detection. The HELLO messages also come with a list of router IDs and a sequence number, such that each node can maintain its neighbor table. The update information is now differential, such that only changes in the router list are transmitted.

In the most recent draft, TBRPF is described as being composed of two main components: neighbor discovery and routing. For routing (as described above) each node computes its source tree, using a modified version of Dijkstra's algorithm. Only significant parts of the source tree are communicated to neighbors. TBRPF also has abandoned the use of sequence numbers, in favor of another technique based on "believing" (trusting) only certain nodes about their topology updates. The draft [117] is very detailed, including precise terminology description, protocol message formats and packet headers and even a detailed algorithm description in pseudocode.

In [12], TBRPF is roughly compared to other ERPF based protocols, but most of them were not designed for a wireless mobile network but for static networks, instead. A simulation based evaluation in this paper compared TBRPF against two slightly different flooding algorithms, but no comparisons against other protocols for mobile ad hoc networks have been done.

[12] is cited just twice and only for example purposes, and also the IETF draft did not seem to be part of any other scientific work on the subject. Apparently there have been no detailed comparisons with TBRPF against any other routing protocol for mobile ad hoc networks.

8.27 TLR/TRR/AGPF - Terminode Routing

Terminode routing is developed at the EPFL in Switzerland. The project aim is to develop a system that is capable of wide area ad hoc routing. The project did explicitly choose an independent roadmap from the IETF MANET working group efforts. Support for IP or interoperability is not the most important requirement, but is not ruled out.

Routing between terminodes is a hybrid process. First the packets are routed based on geographic position. The target address used in this routing is called

LDA (location dependent address). From the target LDA the closest *friend*-node is computed and the packet is passed to it. A *friend* is a selected node in close, but not necessarily direct communication range. If the target node for the packet is among the friends of the node holding the packet, a local routing method is used to pass the packet to its destination.

As position based routing needs some kind of position service, terminodes can use the concept of a *virtual home region* (VHR), which is a *some-for-some* location service approach (cf. [113]). For each node, there exists a "home region", which is fixed and specified by a center position and a radius.Thus the region can be determined by a hash function over the node's id. Each node within the VHR of a certain node must maintain the current position of this node, so that other nodes can obtain it. Thus the home region is in fact independent of the position of its node.

The position-based routing method is called AGPF (anchored path geodesic packet forwarding). As a simple greedy forwarding mechanism doesn't work in many situations (i.e.running into a local minimum), the concept of *anchors* is used. To avoid running into a minimum, the route is oriented on a set of anchors along the path. An anchor is just a specific location independent of any node. The anchored path is determined by the source using FAPD (friend assisted path discovery) and included into the packet (similar

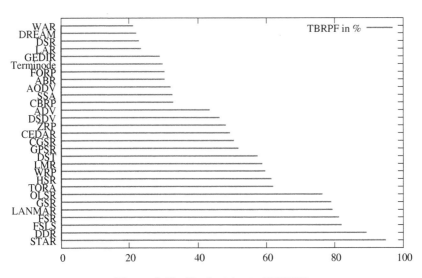

Figure 8.25: Similarities to TBRPF

to source routing).

FAPD is based on small world graphs[47]. Alternatively, the path can be determined by DRD (directed random discovery), which just sends the packet to a set of neighbors whose angle is the smallest to the right direction.

The local routing method is no longer based on position information, but only on a unique node identifier, the *target id*. A two hop neighborhood information is maintained by each node by using HELLO packets. If the neighborhood is known and a packet can utilize local routing (i.e. the target is known to the node which received the packet), a path discovery is initiated to direct the packet to the destination.

The concept of terminodes and terminode routing is described in several papers: [76, 75, 20, 21].

Terminode routing was compared against DSR in simulations, using scenarios which were designed for the use-case of terminode routing (i.e. large areas with large distances, some nodes clustered, with few roaming nodes). In these scenarios terminode routing outperformed DSR by nearly an order of magnitude [21].

The terminode project also addresses some other problems of ad hoc routing. [30] is a paper about positioning without a GPS-like device. The problem of stimulating cooperation of node operators is addressed in [28], which proposes a virtual currency, the *nuglet*: Relaying a message will benefit the relaying node with some units of cash, sending a message to a destination will require some units of cash as "payment". The paper also explains detailed precautions against undesired manipulations.

8.28   TORA - Temporally Ordered Routing Algorithm

TORA is a *link reversal routing* (LRR) algorithm (cf. section 8.22) and was introduced by Park and Corson in [121]. It evolved from LMR and combines also features from Gafni-Bertzekas[55] in a unique single-pass strategy. In this context "single pass" means, that by processing a single event, all route maintenance tasks (erroneous route deletion, search and establishment of new routes) can be combined.

As in LRR algorithms in general, for each destination a destination-rooted DAG is constructed. A height gets associated with each node and thus upstream and downstream links can be identified to route traffic to the destination.

The algorithm itself is rather complex, I refer to the cited literature [121, 40] for a detailed description.

TORA was used in some performance comparisons, notably [25], where it performed very bad. In [40] the authors state, that this is due to the nature of the underlying protocol (IMEP) used in the simulations, which prevents TORA from efficiently using the wireless broadcast channel. Other studies and an analytical comparison against an idealized link state algorithm (ILS) showed excellent performance. [40] also describes an extension which performs a proactive optimization, which may be of use in certain scenarios.

Although TORA can suffer from an unbounded worst-case convergence time, simulations have shown, that even for very stressful scenarios, TORA converges quickly and performs significantly better than the former mentioned ILS algorithm.

8.29 WAR - Witness Aided Routing

Witness Aided Routing by Aron and Gupta[3] is specifically designed to utilize unidirectional links.

WAR makes use of the possibility to overhear any transmission in range of a

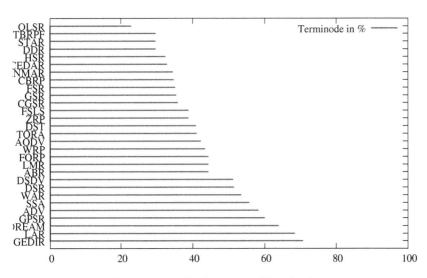

Figure 8.26: Similarities to Terminode

node on a wireless channel in a special way [9]. A node, which can overhear a transmission from one host to another over a relay, acts as a passive *witness* for that transmission. If the relay is not able to reach the destination or does not get an acknowledge, the witness node becomes an *active witness* and tries to deliver the packet on behalf of the relay node, thus saving the packet, even if the original route failed. Because many nodes can be witness of a certain transmission, special care is taken to avoid contention.

The goal is to perform just one single successful delivery. To achieve this, each witness host, which intends to deliver the packet, must get permission from the target host[10]. In order to get the permission, the node sends a request to the target host. If the target host did receive the packet before by the relay (but the witness hosts did not overhear this), the request will be rejected, in any other case, the set of witnesses will be polled by the target until the packet could be successfully delivered.

The route discovery is similar to DSR (cf. section 8.10), with the enhancement of multiple route selection criteria. The target can be instructed to await a certain amount of route requests, or to wait for a certain time period, and then choose the route to answer the route discovery according to

[9]Of course, other protocols, like DSR also make use of that fact.
[10]This is the target host of the witnessed transmission, not the final target of the packet.

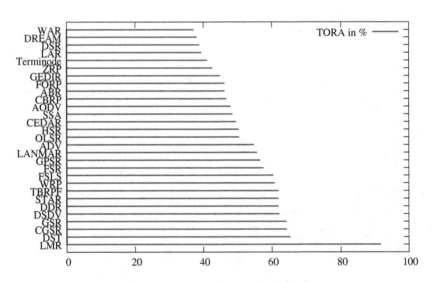

Figure 8.27: Similarities to TORA

some specified criteria. Alternate routes can be remembered, to have them ready if the first choice breaks.

Again like DSR, WAR uses source routing to forward packets. Any forwarding node regards the delivery as successful, if it receives an acknowledgment from either the intended relay node or from any witness. If not, the route is considered broken and a route recovery process is initiated. Just like DSR the source route information in a relayed packet can be used to update local routing information.

Route recovery works by broadcasting the packet to all neighbors of the host, which failed to deliver it to the next hop, and setting a special flag. These hosts now try to deliver it, using the remaining source route information, treat it as a regular packet and clear this flag. However a packet can only be recovered a fixed number of times, which is set by the source. If all these attempts fail, the acknowledge for the packet will eventually time out at the source and the source will reinitiate a route discovery.

[3] gives a short analytical comparison against DSR and provides constraints under which WAR is more bandwidth efficient, than DSR. A much more detailed analytical study is presented in [4]. This study aims to prove the scalability of methods like WAR and problems with on-demand routing protocols like DSR. [5] is the master thesis by Ionut Aron about the subject, which provides the same results in more detail.

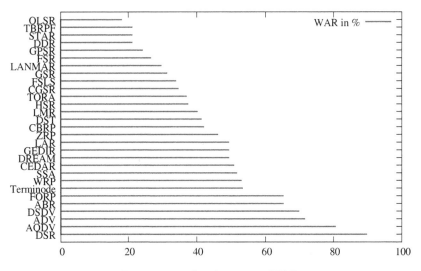

Figure 8.28: Similarities to WAR

8.30   WRP - Wireless Routing Protocol

The Wireless Routing Protocol by Shree Murthy and J.J. Garcia-Luna-Aceves is one of the first suggestions of a routing algorithm for mobile ad hoc networks. It was proposed 1996 in [114] and the only other protocol mentioned therein is DSDV (cf. section 8.9).

WRP is related to the DBF[14] algorithm. Routing update messages are only sent locally to the neighbor set. They contain all the routing information the originating node knows of. Of course not the whole routing table is sent in each update. Only changes are transmitted, either by receiving an update from another node, or of a link in the neighborhood has changed. WRP is a proactive routing protocol, since routes are maintained all the time and no special route requests by source nodes need to be performed.

The routing table consists of an entry for each destination with the next hop and a cost metric. The routes are selected by choosing the node from the neighbor set, which provides the path with the lowest cost (provided it's loop-free), as next hop. The link costs will be kept in a separate table, but it is not specified, how the cost for each link should be determined. Various possibilities exist: hop count, end-to-end delay, utilization, etc.

To keep the state of the neighbor links up to date, empty update messages (HELLO messages) are sent in a regular fashion, if no other updates would be sent anyway. Update messages which are not empty, need to be acknowledged.

[114] presents a proof of correctness and some simulation results, where WRP is compared against DBF, DUAL (the routing algorithm from EIGRP) and ILS (and idealized link state algorithm). The simulation was very simplified, a few simple static topologies have been simulated with randomly forced link failures to model movement related link breaks. The message overhead is counted for the regarded protocols. The results are clearly in favor of WRP. WRP is referred to in a lot of papers, mainly due to the fact, that is one of the earliest proposals. The authors continued some work on WRP, which lead to WRP-lite in [132], which is later called BEST (Bandwidth Efficient Source Tracing) in [133][11].

8.31   ZRP - Zone Routing Protocol

The Zone Routing Protocol by Zygmund Haas was first introduced in [67].

---

[11]Among others, BEST and DST (Distributed Source Tracing) are not further discussed in this study.

It is a hybrid protocol, that combines reactive and proactive strategies. Since the advantages of either approach depend on the characteristics of the network (like the degree of mobility), it could be beneficial to combine them.

ZRP introduces the notion of a *routing zone*, which is a set of nodes within the local neighborhood. In practice the zone is defined by the maximum number of hops, a node within the zone may be distant from the zone's center node. Each node maintains routing information actively within its zone. The algorithm used is called *Intrazone Routing Protocol, IARP*. A basic link state algorithm is used for this purpose.

To discover a route outside the local routing zone, a reactive protocol, the *Interzone Routing Protocol, IERP* is used. For this purpose a *bordercast* of a request message is used. *Bordercast* means, the request is forwarded to the peripheral nodes of the zone, which in turn can check if the target is within their own zone, or continue to bordercast (cf. Section A). The bordercast process must take care, not to bordercast requests back into regions already covered. To achieve this, queries must be recorded for some time by the relaying nodes. ZRP uses a special technique for this, called *Advanced Query Detection* and *Early Termination*. Route caching and local repair is also possible.

Additionally to [67], ZRP is described more detailed in [69]. Some more

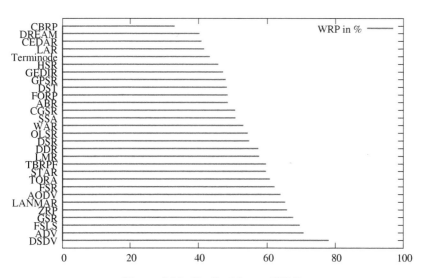

Figure 8.29: Similarities to WRP

investigations have been published in [68]. ZRP is also described in [70]. ZRP was also mentioned as a reference protocol that utilizes the hybrid approach. However it was not used in independent performance comparisons.

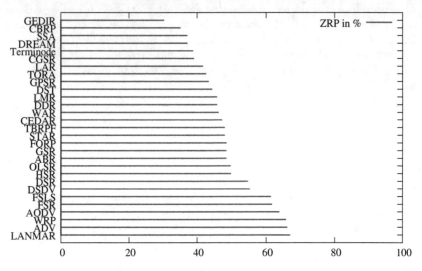

Figure 8.30: Similarities to ZRP

Chapter 9

Realistic Scenarios for Evaluation

In this chapter, I will propose requirements for good evaluations and introduce some tools to help fulfilling some of these requirements. The main goals for my understanding of good quality scenarios is realism together with applicability for the intended use cases of the scenario.

9.1   Requirements of Scenarios

From the previous chapters 4, 5 and 6, some obvious points have emerged, that could lead to a significant improvement in the applicability of simulation based evaluation of MANET routing protocols.

Most promising is the use of obstacles in evaluations. This is backed by the clear result, that the evaluations done by Per Johanssen et.al. in [85] are far superior than most other evaluations. Obstacles will serve two purposes: they influence the movement of the nodes and they obstruct connectivity. While the blocking of radio waves by obstacles needs support of available simulation software, it is possible to use the altered movement patterns and better mobility models even with current simulation software. A way of using obstacles was just recently discussed in [81], which proposes an *Obstacle Mobility Model*. This model uses the edges of the Voronoi diagram of the area (including obstacle shapes) as pathways for nodes. This paper actually shows how to include obstructive behavior of obstacles to the transmissions with only minor interference with the simulation software. In this case a reachability matrix is computed for each step, and a packet which should not be received according to the matrix is dropped by the receiver.

However, not just obstacles can lead to a more realistic behavior of nodes, but as well turning and acceleration, or more general, correlated movement (cf. section 6.3).

The following section describes guidelines, how to take these things into account. The concept of obstacles is further generalized to the concept of a *region* or *subarea* which can influence directly nodes within or near that area. The forced concentration of nodes due to certain constraints (like traffic lights, cars on street, areas of attraction) is an effect that is likely to have a significant impact on the performance of an ad hoc network.

I will propose an example, how to formally describe a scenario, that can take many of the interacting properties into account (although it cannot be comprehensive). The following model will focus on the *strategy*, and has the option to include a detailed *kinetical state* model, but this is not required.

9.1.1   The Node-Interactive Mobility Model

I will introduce a mobility model based upon several components that can be described through various parameters. An implementation needs to take these parameters into account and create movement instructions for the nodes according to these specified rules. The main aspect is that nodes can interact with their environment, hence the name.

Basic Components and Types

Nodes are the basic components of the simulation and they are also basic components of the scenario description. As a scenario likely includes various different types of nodes, I allow to define these types.

A node type definition should include the following characteristics:

- maximum speed (mandatory)

- maximum acceleration (optional)

- maximum deceleration (optional)

- turning factor[1] (optional)

Nodes may further be combined into groups, that will show some common movement behavior. The group may act as a large blurred node with typical node characteristics as described above. Additionally the group will have characteristics like:

---

[1]This determines, how fast the node is able turn at a certain velocity.

- maximum diameter

- node movement strategy within the group

- node density within group

- number and types of group member nodes

- probability of nodes joining or leaving the group

Then it is required to define certain types of **regions** within the area. The regions itself are described by geometric properties and can be of a certain type. The type of a region determines, how the region interacts with the node types.

A possible set of characteristics of a type of a region is:

- attractive to nodes of types [..] by degree $d \in [0..1]$

- forbidden for node types [..]

- restricted to node types [..]

- obstructing radio waves to a degree of [..]

- maximum speed in this region limited to [..]

- some characteristics are only valid during certain time intervals[2]

A value greater than 0.5 for the *degree of attraction d*, is considered attractive, a value below is repelling, exact 0.5 would be neutral. This property can also be included into the nodes, such that nodes of certain type are attractive or repellent to each other (e.g. cars like to avoid other nodes to prevent an accident, so they would be repellent to any other node to a certain degree).

Scenario Specification

Now from the defined types I can build the specification, which needs to contain:

- Set of n nodes of type T_n

- ...

---

[2]This can be used to model traffic lights at an intersection.

- Nodegroup containing x nodes of type T_n'

- ...

- Region of type T_r, at coordinates $(x1, y1, x2, y2, \ldots)$

- ...

These specifications should define a scenario with individual movement strategies for nodes and region and node dependent constrains. This allows to derive a concrete scenario with exactly determined node movements, suitable to feed into a simulator. The specification leaves enough freedom for sensible random behavior in terms of probability functions, without letting the nodes just behave in an arbitrary manner.

Formal Specification Aspects

For an implementation, the scenario specification described needs to be available in a machine readable format. I suggest to use XML[165] for this purpose, since parsers and validators are widely available. XML is flexible enough to allow specification of all aspects in a simple and intuitive way. The specification can easily be extended or can include data, which may not yet be used by the application. Other specification languages, e.g. C-style are also possible. The implementation of a scenario generator according to these guidelines is described in the following section 9.2. A sample configuration file is given in appendix D.

9.2   Partial Implementation of the Node-Interactive Model

In order to make use of the proposed guidelines, a student project was issued and supervised by myself. The assignment covered the implementation of a scenario generator, that takes the described characteristics of a good scenario into account as far as possible. The resulting work was done by Jan Lange and is described in his thesis [97]. This section will briefly describe the achievements of our work.

9.2.1   Requirements

The main requirements of the student project implementing such a scenario generator were divided into two groups:

Requirements on the specification of a scenario

It should be possible to specify the following characteristics:

- Different types of nodes in a scenario and their characteristics can be specified independently.

- A node-type should be able to move with varying velocity. Changes should be performed by acceleration and deceleration.

- Different nodes should be able to use a different movement strategy. A main movement target or direction should be result of such a strategy, but it should still be possible for nodes to frequently change directions and speed.

- Node types should be able to decide upon direction changes dependent on other nodes, while a general strategy should be independent.

- Area types and areas should be defined, that could modify the characteristics of nodes internally, depending on their types.

- Areas could be restricted to all nodes, certain types of nodes or no nodes at all.

- Areas should be able to attract or repel nodes.

Requirements on the implementation of the generator

- There should be a sensible way to specify the scenario, e.g. a descriptive parameter file.

- The scenario generator should be easy to use.

- The implementation should be stable, robust (particularly against wrong input data) and correct.

- The scenario generator should scale well, even with a very large number of specified items.

- The implementation should be modular and provide a clearly defined interface in order to add future functionality and enhancements easily.

- The output should result in sets of files, that could be used by common simulation software as input.

- The implementation should be platform independent and run on the most common systems.

9.2.2   Outline of the Implementation

The resulting scenario generator was implemented in Java(tm)[83], which yields platform independence, object-oriented structuring for modular programming and later enhancements. The design was strictly modular with clear interfaces to allow interchangeable modules for the various components. The design is described in figure 9.1, adapted from [97].

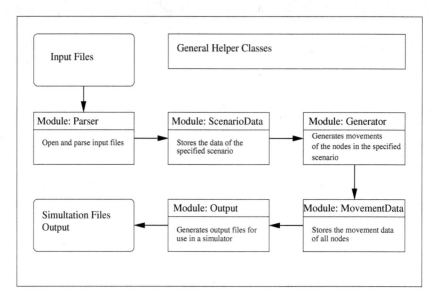

Figure 9.1: Structure of the scenario generator implementation

The input data is read from a file in XML[165] format, providing data for the scenario specification like node types and characteristics, area size, etc and also general parameters like output filenames and random seed. A sample configuration file is given in appendix D.

The application is command-line driven, although a GUI which visualizes progress can be enabled optionally. The command line interface allows the use of the generator in a scripting framework and to schedule a batch of generating jobs in a simple way and even distribute jobs on multiple hosts for parallel generation of scenarios.

The modular design allows to use different output back-ends, to generate output in different formats. Until completion only one output module was implemented, which produces output that can directly be used by NS-2[116] tcl-scripts.

The main components, that can be specified in a scenario are the following:

- Node types with different movement characteristics, including maximum speed, acceleration, likelihood of direction changes and attraction to other node types.

- Group types, which are defined by a single node type (all nodes in the group are of this type) and a maximum distance, that is allowed for each node from the center of the group.

- Area types, which can have the following characteristics: blocking of radio waves (no effect, since simulation software does not support it), the area is prohibited for certain node types, attraction of certain node types which can also be effective only during certain time periods. Further, the area can change the general characteristics of all nodes within.

- Then the actual objects, i.e. amount of nodes of a certain node type (or group type) and initial location, areas of a certain type within given positions and areas can also contain points of attraction, which act as focal points for nodes, that are attracted by this area.

- Finally general information, like the overall dimensions of the scenario area, the duration of the experiment and the number of all nodes.

Generator Algorithm

The algorithm to calculate the movements will be outlined in this subsection. In order to provide good scalability an event-driven approach was used. Thus it is not required to recalculate every movement for every node every certain time interval. Instead events are inserted in a priority queue, which is ordered by the simulation time. The basic event is the calculation of the next move of each node. It can be predicted, when the move is completed and the next calculation event needs to take place. This following event is inserted into the queue. Other events happen, if a node enters or leaves an area, or when an area changes the characteristics of nodes periodically. Since nodes can not collide with each other, there is no event, that would alter the precalculated next event for any node.

Several aspects influence the movement decision of each node. Nodes in the neighborhood and areas (with or without focal points) can attract the node or make him leave the current area on the shortest path. Further each node has some additional random way to choose its direction. The movement direction is computed from these attracted directions.

Since neighbor nodes affect the movement of a node, the neighbor set must be computed. The straightforward way would be to calculate the distances of all other nodes to the node and choose those that are within a certain range. This would impose a very high computing complexity for each movement calculation. Thus another approach was chosen. The simulation area is divided into "fields" by using a grid. Each field keeps track of nodes, that are within. This results in another event, if a node changes it's current field. Neighbors are simply nodes that are within the neighboring fields. The field size (grid width) can be configured.

9.2.3   Limitations of the Implementation

Since the scenario generator was implemented in a student project in a rather short time, not all requirements and desired features have been met. There is just one back-end for the NS-2 simulator, back-ends for GloMoSim/QualNet or other simulators are still missing. The implementation uses a rather course grain time resolution of one second. To change this to a more fine grained resolution, would require some rework of the implementation. Finally, the nodes choose their own initial direction in a random manner. A strategy module would be a good addition, although this initial direction is modified by potential points of attraction (areas or other nodes).

9.2.4   Achievements of the Implementation

Although there are some limitations, the implementation of the scenario generator addresses many of the problems with past scenarios. The resulting scenarios appear much more realistic than the scenarios used in previous evaluations. Further the design of the generator allows to create a large batch of scenarios for simulation, and is more easy to use than GUI driven generators like CADHOC[142]. Since the project was very time limited and had to be completed in four months, the result is quite acceptable.

Chapter 10

Simulation Framework

This chapter describes a framework to facilitate simulations to evaluate mobile ad hoc network routing protocols.

10.1 Motivation

I already described some problems with simulations based evaluations of mobile ad hoc network routing protocols, like non-representative simulations scenarios. However, there are more problems. There are only few papers that evaluate more than one or two routing protocols. In most cases the currently proposed routing protocol is compared to one or two already established or well-known protocols. There are rare exceptions like [25] that compare more routing protocols without proposing a new protocol (although some of the authors are indeed the developers of one of the compared protocols).

Still, a broader comparison would still lead to much more insight, which protocols are best appropriate for what kind of application scenarios. In order to achieve such a comparison a large number of simulations would be required.

10.1.1 On the Amount of Simulations to be Performed

In order to judge the effort required to perform comprehensive studies, it is inevitable to examine the expected amount of simulations, that are necessary to yield useful results.

There are several factors that come together:

- Simulation Scenarios

- Routing Protocols

- Repetition of various steps, in order to get good statistical results.

Simulation Scenarios

I have proposed eleven different application scenarios, each of which can be varied again to some extent. I assume, it is sensible to evaluate five variants of each application scenario on average, resulting in 55 scenarios.

Since not all scenarios apply to all routing protocols, it is not required to simulate any combination, but I assume that at least six (of eleven) major application scenarios apply to each routing protocol. Thus 6 (application types) $\times 5$ (variants) $= 30$ (scenario specifications) would need to be simulated with each routing protocol to examine.

Routing Protocols

I have described and classified 30 different routing protocols and meanwhile it is possible that new have been developed and further I am aware, that I have omitted some.

However, in section 7.4, I have defined a set of functions to yield a similarity relationship between the routing protocols. This relationship can be used to reduce the number of routing protocols to examine in a study. Not all routing protocols need to be tested, but only a representative protocol for a certain class, provided the protocols in such a class behave similar enough under the given conditions. For example in figure 7.4 TBRPF, STAR, DDR and GSR are closely related and thus can be considered to form a class. One could pick just one of these protocols as a representative and just evaluate the scenarios with this one. In this example, one would choose TBRPF, since it is developed most mature compared to the others.

The question remains, if the similarity relationship, I have presented is appropriate. This also must be confirmed by simulation, such that it is confirmed or denied, that those routing protocols that are closely related in terms of my standards do indeed behave similar under the given conditions. Most likely this will not be entirely the case, but in an iterative process, the similarity functions can be tuned to yield an appropriate similarity relationship. For these simulations to tune the similarity functions, again scenarios would be required. Which somehow leads to a hen-and-egg problem. Regardless, the similarities identified in section 7.4 should provide a useful start. In figure 7.4 six classes remain, however many routing protocols are no longer listed in the figure since they have no relationship strong enough. So they would either form classes with a weaker relationship or would form single classes. Assumed we would put them into a singleton class each, this would result in

13 additional classes with 19 total.

Statistical Repetition

Currently, the amount of simulations to be performed would be 30 scenario instances to be simulated with at most 19 routing protocols, which yields 570 simulation experiments. While this is already a large number, it is not unfeasable.

However, this number is not correct. In order to get good results and to rule out statistical coincidences, each major step which uses random data, has to be repeated. I consider the generation of a simulation scenario from a scenario specification and the actual simulation with simulation software as such steps.

Thus, from each of the 30 scenario specifications a set S_n has to be generated. Thus the amount of scenarios to be simulated is:

$$|S| = \sum_{i=1}^{30} |S_i| \tag{10.1}$$

The same applies to the actual simulations. For each generated scenario $s \in S$ a set of experiments E_s needs to be performed.
The total number of experiments is consequently:

$$|E| \;=\; \sum_{j=1}^{19} \sum_{s \in S} |E_s| \tag{10.2}$$

$$\approx \;\; 19 \cdot |S| \cdot |\bar{E}_s| \tag{10.3}$$

$$\approx \;\; 19 \cdot 30 \cdot |\bar{S}_n| * |\bar{E}_s| \tag{10.4}$$

$$\tag{10.5}$$

The values of $|S_n|$ (amount of scenarios to generate from a specification) and $|E_s|$ (amount of experiments to perform with each scenario and routing protocol) cannot be defined to some fixed values.

Instead they depend on the desired confidence of the statistical results. Any required amount of experiments can be determined depending on the desired confidence interval as follows:

$$N = \frac{s \cdot t_{1-\alpha}(f)}{|CI|} \tag{10.6}$$

with s being the estimated standard deviation, t being the *Student's t-Distribution*, α the probability of error, f the amount of degrees of freedom and $|CI|$ the size of the confidence interval.

To determine N, an iterative process is required, because the standard deviation can only be estimated in advance. An initial set of experiments has to be done. From these initial results, it can be computed if more experiments are required to fulfill the desired constraints.

There was only one notable paper [36] that documented and justified their amount of experiments (30 in their case).

Resulting Amount

If the number 30 would serve as an example value for $|\bar{S}_n|$ and $|\bar{E}_s|$, the total amount of simulation experiments to be performed would be $19 \cdot 30 \cdot 30 \cdot 30$, resulting in 513000 experiments to be set up and run in total.

Considering that a successful simulation run using GloMoSim (however, to get successful runs proved to be a major problem, cf. section 11.7.1) on a Sun Fire V880 machine takes on average 7.74 hours (cf. section 11.7.2), the total calculation time would exceed 450 years.

Even if this is an exaggerated number and the problem can be parallelized very well, the number of scenarios to be generated and simulated will still be very large. Therefore it is inevitable that the generation of scenarios and the simulation of these scenarios is as much automated as possible. For instance, it would even be nearly impossible to create a few hundred simulation scenarios, if this would have to be done by hand using a GUI driven tool. But even with a scriptable scenario generator, there are many steps from the specification to the simulation:

- Specification of the movement scenario parameters.

- Specification of the traffic scenario parameters.

- Specification of the simulation parameters (e.g. duration, radio parameters, etc)

- Generation of the scenario.

- Conversion of the scenario data into a format for the simulator.

- Generation of the simulator configuration and input.

- Running the simulator.

- Process simulator output for result data.

- Interpret result data.

Some of these steps are minor, some can be very complex (e.g. processing simulator output is a major problem if standard NS-2 is used, due to the huge amount of trace-file data). In a real simulation project, additional steps are required, e.g. for distributing simulation batch jobs on a set of computers for parallel simulations.

10.2 Proposed Solution

Many of these steps can be automated if an appropriate framework is used. This will enable us to set up and perform simulations much more efficient and consequently more simulation experiments can be performed yielding better and more comparative results.

I have assigned another student project to design and implement such a framework to facilitate efficient simulation for evaluation of mobile ad hoc routing protocols.

10.2.1 Requirements on the Implementation

The assignment included the design and implementation of a simulation framework, that meets the following requirements:

- To minimize the manual effort required to perform simulations.

- Capability to generate a set of similar but different scenarios to a single scenario specification (by setting a different random seed).

- Support multiple movement and traffic scenario generators and provision of interfaces to facilitate the use of future movement and scenario generators.

- Support for multiple simulation software packages and provision of interfaces to facilitate the use of future simulation software packages.

- Platform independence.

- Good documentation to allow future enhancements of the framework.

- Provide post-processing of simulator results and display processed results in a common and clear way.

10.2.2   Outline of the Implementation

The implementation of the project resulted in the simulation framework called "SimFrame" and is described in detail in the coursework of the authors Levin Brunner and Hannes Schmidpeter [26].

Their simulation framework is a modular program implemented in *perl* [162]. It supports two scenario generators: ANSIM[72] and NoMoGen, which is the name of the scenario generator developed by Jan Lange [97] in an earlier project (cf. section 9.2).

It supports the NS-2 simulation software [116] and provides basic interfaces to include support for other simulators. As traffic generator only the internal mechanisms of NS-2 can be used, which allow to create traffic in the simulation. But a modular interface for other traffic generators to be plugged-in, is available.

It's operation works as follows: The scenario specification (as well as all other configuration data, like pathnames, simulator used, etc) is read from an input file in XML format. The input is parsed and the data is passed to the each corresponding module.

The movement scenario specification is passed to the module that handles the movement generator. This module creates an input file and command lines for the configured movement generator (e.g. ANSIM) and calls it. The same happens to the traffic scenario specification. The resulting data from the scenario generators is transformed to be used by the target simulation software. Further general parameters for the simulation software are assembled and processed to yield all necessary data including command lines, to launch the simulation. A script will be created, that performs the specified simulations. It can be started automatically or manually. Post-processing of the results is not performed by the framework, yet. The architecture of the simulation framework is described in figure 10.1.

10.2.3   Limitations of the Implementation

Current only NS-2 is supported, this is a severe limitation. Another big limitation is, that there is no post-processing performed. This requirement was canceled, since trace-file processing of NS-2 was considered the wrong way. Rather NS-2 should be extended to provide reasonable statistic data if required, but this was beyond the scope of the student project.

10.2.4 Achievements of the Implementation

The simulation framework can automate many tasks to perform simulations with NS-2 as long as one of the supported scenario generators is available.

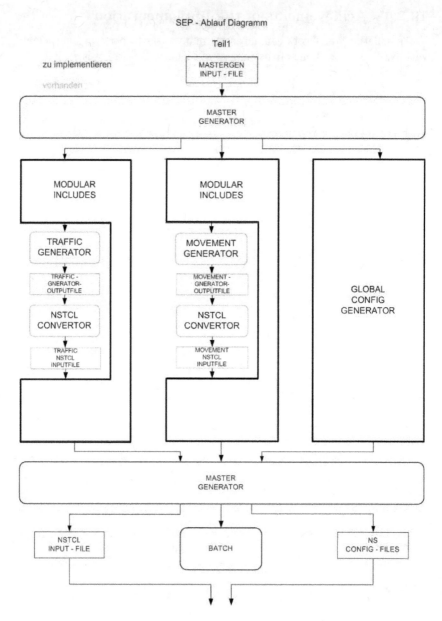

Figure 10.1: Architecture of the Simulation Framework

Chapter 11

Simulations

The following chapter describes which simulations have been performend and their results.

11.1   Aims of these Simulations

These simulations were performed to examine (on a sample basis only) the following questions and issues raised by this research work:

1. Confirmation of similarity of scenarios, i.e. similar scenarios will yield a similar result for the same routing protocol.

2. Do realistic (and more complex) scenarios yield much different results than simple scenarios in terms of which routing protocols would perform best? (In order to examine this question, realistic scenarios were created using the NoMoGen generator and simple random waypoint scenarios were created with the scengen generator.)

3. Confirmation of similarity of routing protocols, i.e. similar routing protocols will yield a similar result for the same scenario. (Do routing protocols yield similar results in simulations, if they are considered *similar*, as defined in section 7.4).

4. Evaluation of routing protocols in different movement and traffic scenarios, using selected scenarios typically for certain classes of intended applications. I.e. which of the examined routing protocols yield best results in a particular scenario. (This is the basic question that is behind all these anyway and although there may not be a definite answer, yet, there is no reason not to ask this question.)

5. Last not least, provide better understanding of the special problems concerning simulation based evaluation of routing protocols and their scenarios.

11.2 Performance Metrics and Sample Statistics

Ideally the performance metrics as described in section 2.1.1 would be measured in the same way for all simulation experiments. Unfortunately GloMoSim does not allow to collect all these values in a common and consistent way. Instead only the following have been available for measurment: (For all ratios, the values were only taken into account if they were defined, i.e. the divisor was > 0.)

11.2.1 Routing Overhead

This is the ratio of routing control packets to all IP packets being sent (including broadcast packets). Since all concerned routing protocols are using IP packets (either directly or indirectly), this ratio gives the routing protocol overhead in terms of packets. The routing control packets are measured differently depending on the protocol. The best parameter provided by GloMoSim to measure the IP Packets transmitted is:

NetworkIp, Number of Packet Attepted[1] to be Sent to MAC

which is common for all routing protocols. However for the number of routing control packets, there is no common result parameter available. So the following values have been used.

AODV and DSR

For AODV and DSR, there is a value that can be used directly.

GloMoSim result parameters used:

- *RoutingAodv, Number of CTRL Packets Txed (n_{rp})*

- *NetworkIp, Number of Packet Attepted to be Sent to MAC (n_{ip})*

---

[1]This is actually the exact way how the result parameter is produced by GloMoSim, although it would be correct to write ...Packets Attempted... instead of Packet Attepted

Routing overhead calculation:

$$r_{ro} = \frac{n_{rp}}{n_{ip}}$$

WRP

For WRP, the only values available are *RoutingWrp, Number of routing packets sent* and *RoutingWrp, Number of routing packets recvd*. There is no difference made between route request or route replies because WRP only sends proactive routing updates. To calculate the overhead I use the number of routing packets sent.

GloMoSim result parameters used:

- *RoutingWrp, Number of routing packets sent* (n_{rp})
- *NetworkIp, Number of Packet Attepted to be Sent to MAC* (n_{ip})

Routing overhead calculation:

$$r_{ro} = \frac{n_{rp}}{n_{ip}}$$

LAR1

For LAR1 a lot more detailed values are available. Those representing route requests, route replies and route errors sent are summed up to get the total number of routing control packets.

GloMoSim result parameters used:

- *RoutingLar1, Route Requests Sent As Data Source* (n_{rreq})
- *RoutingLar1, Route Replies Sent as Data Receiver* (n_{rrep})
- *RoutingLar1, Route Error Packets Sent As Source of Error* (n_{rerr})
- *NetworkIp, Number of Packet Attepted to be Sent to MAC* (n_{ip})

Routing overhead calculation:

$$r_{ro} = \frac{n_{rreq} + n_{rrep} + n_{rerr}}{n_{ip}}$$

FSR

For FSR only the number of intra and inter scope updates are counted. Although there is also a value named "packets received from UDP", this value does not seem to make any sense. I assume it is a number in bytes instead of packets. Again, the number of intra and inter scope updates is divided by the number of IP packets sent to yield the routing overhead.

GloMoSim result parameters used:

- *RoutingFisheye, The number of Intra Scope Updates* (n_{intra})

- *RoutingFisheye, The number of Inter Scope Updates* (n_{inter})

- *NetworkIp, Number of Packet Attepted to be Sent to MAC* (n_{ip})

Routing overhead calculation:

$$r_{ro} = \frac{n_{intra} + n_{inter}}{n_{ip}}$$

11.2.2   Route Setup Ratio

The route setup ratio should be the amount of route replies received divided by the amount of route requests sent. "Should be" because, alas, GloMoSim does not provide these values.

AODV, DSR and LAR1 code provide route requests sent and route replies sent (but not route replies received), which makes a big difference. FSR, even worse, does not provide any such values, thus always being displayed with a ratio of 0. WRP does not report requests and replies separately (and proactively updates all routes), thus for WRP the ratio between sent and received routing packets has been used (although it is not really the same route setup ratio but the ratio of successfully transmitted routing updates). Although most such values are not exactly what would be required, I still decided to display these values, since even such rough values can be interpreted and will help to explain a certain behaviour.

AODV GloMoSim result parameters and calculation:

- *RoutingAodv, Number of Route Requests Txed* (n_{rreq})

- *RoutingAodv, Number of Replies Txed* (n_{rrep})

$$r_{rsr} = \frac{n_{rrep}}{n_{rreq}}$$

DSR GloMoSim result parameters and calculation:

- *RoutingDsr, Number of Requests Txed (n_{rreq})*
- *RoutingDsr, Number of Replies Txed (n_{rrep})*

$$r_{rsr} = \frac{n_{rrep}}{n_{rreq}}$$

LAR1 GloMoSim result parameters and calculation:

- *RoutingLar1, Route Requests Sent As Data Source (n_{rreq})*
- *RoutingLar1, Route Replies Sent as Data Receiver (n_{rrep})*

$$r_{rsr} = \frac{n_{rrep}}{n_{rreq}}$$

WRP GloMoSim result parameters and calculation:

- *RoutingWrp, Number of routing packets sent (n_{sent})*
- *RoutingWrp, Number of routing packets recvd (n_{recv})*

$$r_{rsr} = \frac{n_{sent}}{n_{recv}}$$

Nothing could be measured for FSR.

11.2.3 MAC Broadcast Ratio

The is the simple ratio between MAC broadcast packets sent (as registered by sending nodes) and MAC broadcast packets received. This ratio is basically the "multiplying factor" of mac broadcasts, since a broadcast packet sent is always only counted once (at the transmitting node) but received multiple times (at all receiving nodes). Two GloMoSim provided result values can be used directly and for all routing protocols, since this value on the link layer is of course independent of any routing protocol code.

GloMoSim result parameters and calculation:

- *802.11, BCAST pkts rcvd clearly (n_{recv})*

- *802.11, BCAST pkts sent to chanl (n_{sent})*

$$r_{mbcast} = \frac{n_{recv}}{n_{sent}}$$

11.2.4   IP Delivery Ratio

This is the ratio of IP packets sent to the MAC layer (either as originating node or as intermediate node routing the packet to its next hop) and IP packets received from the MAC layer (either as destination or as intermediate node). However, all routing control packets cause the *IP packets sent to MAC* counter to increase (since they use the IP layer in GloMoSim), but the "received counter" is not increased for routing protocol data (except for WRP and FSR, because they utilize UDP and UDP packets are counted as received). Thus, this dividend of this ratio is adjusted by the number of routing control packets to correctly handle this variation.

Further since FSR and WRP utilize UDP broadcast packets, it happens for each broadcast packet, that it is received by many nodes. However, it is not possible to filter these packets from the resulting statistics, thus leading to an IP delivery ratio sometimes $>> 100\%$. This is not the case for the other protocols, since routing protocol IP packets are not counted at all in the IP receiving module. Unfortunately it is not possible to distinguish the broadcast UDP packets from Unicast ones. However, it is possible on the MAC layer, thus looking also at the MAC broadcast ratio will allow to put such results more into perspective.

WRP and FSR GloMoSim result parameters and calculation:

- *NetworkIp, Number of Packets Delivered To this Node (n_{recv})*

- *NetworkIp, Number of Packets Routed For Another Node (n_{rout})*

- *NetworkIp, Number of Packet Attempted to be Sent to MAC (n_{sent})*

$$r_{ip} = \frac{n_{recv} + n_{rout}}{n_{sent}}$$

AODV, DSR, LAR1 GloMoSim result parameters and calculation:

- *NetworkIp, Number of Packets Delivered To this Node* (n_{recv})

- *NetworkIp, Number of Packets Routed For Another Node* (n_{rout})

- *NetworkIp, Number of Packet Attepted to be Sent to MAC* (n_{sent})

- Routing Control Packets (n_{rp}), as used as dividend in the Routing Overhead calculation (cf. section 11.2.1).

$$r_{ip} = \frac{n_{recv} + n_{rout} + n_{rp}}{n_{sent}}$$

11.2.5 Broken Links/Robustness

This is an indicator of the robustness of the protocol. Unfortunately only AODV and DSR simulations report the number of broken links in active routes. FSR, WRP and LAR1 do not report this.

AODV and DSR GloMoSim result parameters and calculation:

- AODV: *RoutingAodv, Number of Broken Links*

- DSR: *RoutingDsr, Number of Link Breaks*

These value are used directly.

11.2.6 TCP Retransmitted Packets

The number of TCP packets that needed to be retransmitted. This is again an indicator of robustness, like "broken links", but fortunately it is available for all simulations.

GloMoSim result parameters and calculation:

- *TransportTcp, data packets retransmitted*

This value is used directly.

11.2.7   Application Data Throughput

Because the applications do not try to maximize throughput, this is a value
of limited usefulness (it does not say anything about maximum achievable
throughput, because the applications did not attempt to maximise the throughput). Escpecially TELNET is usually a low bandwith application with low
throughput. A more valuable measure would have been the end-to-end delay
for telnet application, but unfortunately this is not available. For FTP it
is more useful, because FTP usually tries to transfer the files as quickly as
possible.

GloMoSim result parameters and calculation:

- *AppTelnetServer* result line provides values (including throughput) for
  each telnet connection. The throughput of an individual telnet connection i is T_{tc_i}. The number of individual telnet connections is n_{tc}

- The same applies to *AppFtpServer* for FTP connections. The throughput of an individual FTP connection i is T_{fc_i}. The number of individual
  FTP connections is n_{fc}

- The same applies to *AppHttpServer* for HTTP connections. The throughput of an individual HTTP connection i is T_{hc_i}. The number of individual HTTP connections is n_{hc}

The calculation for TELNET throughput:

$$T_{tavg} = \frac{\sum_{i=1}^{n_{tc}} T_{tc_i}}{n_{tc}}$$

The calculation for FTP throughput:

$$T_{favg} = \frac{\sum_{i=1}^{n_{fc}} T_{fc_i}}{n_{fc}}$$

The calculation for HTTP throughput:

$$T_{havg} = \frac{\sum_{i=1}^{n_{hc}} T_{hc_i}}{n_{hc}}$$

11.2.8   TCP Delay

Mean TCP delay can be computed by using the accumulated application
throughput and also the number of bytes transmitted, which is also available
from the application result values.

GloMoSim result parameters and calculation:

- The throughput values as in the previous section.

- Summarized throughput in bytes per second (using the results as described above) d_{bps}

- Total number of bytes sent n_b again summed up from the *AppTelnetServer*, *AppFtpServer* and *AppHttpServer* result lines.

In addition I assume there is a mean packet size $\bar{s}$, a mean number of packets sent: $\bar{f_p} = \frac{n_b}{\bar{s}}$ and a mean throughput in packets per second $\overline{d_{pps}} = \frac{d_{bps}}{\bar{s}}$. The mean delay of the data packets is thus:

$$\bar{y} = \frac{\bar{f_p}}{\overline{d_{pps}}} = \frac{n_b}{\bar{s}} \cdot \frac{\bar{s}}{d_{bps}} = \frac{n_b}{d_{bps}}$$

with:

$$d_{bps} = \sum_{i=1}^{n_{tc}} T_{tc_i} + \sum_{i=1}^{n_{fc}} T_{fc_i} + \sum_{i=1}^{n_{hc}} T_{hc_i}$$

11.3 Simulation Tools Used

In order to perform useful simulations to evaluate mobile ad hoc networking routing protocols in selected scenarios in an efficient manner, the toolset (scenario generator, simulation software, scripts, etc.) must meet a set of requirements. These are outlined in this section.

11.3.1 Requirements

The ideal requirements to such a package are now summarized, to help choosing the best possible simulation software package available:

- It should be possible to specify the movement scenario, traffic scenario, routing protocol and the desired result values to be measured independently in a clear way and it should be well documented, how to do this.

- There should be an (expandable) library of routing protocols, all following a strict standard to take a standard set of input parameters and to provide a standard set of output values as above (of course there can be additional specific input parameters). The initial library should ideally cover the most discussed routing protocols of each class, e.g.: AODV, DSR, FSLS, FSR, LAR, OLSR, STAR, Terminode, TBRPF, ZRP.

- There should be an (expandable) library of sample movement scenarios, ideally representing each an intended application of MANETs.

- It should be possible to specify traffic scenarios in a detailed way (e.g. using a recorded network trace), as well as to use a a stochastic generator (which only needs some initial parameters set up) to create traffic patterns during the simulation. The generator should record the resulting traffic in a network trace suitable for future input to the simulator. This is important for a strict comparison of routing protocol to rule out stochastic effects resulting from individually generated traffic. A library of sample traffic traces should be provided (again they should represent intended applications for MANETs).

- It should be possible to place obstacles into the simulation area, which interfere with the radio communication as well as with movement.

- There should be run-time control mechanisms to suspend and save current simulation state for later resume, automatic checkpoint creation and event triggers (that print certain values upon receiving event notification). This would be very useful for debugging and to detect statistical effects like the end of an unstable startup phase.

- Ability to perform long duration experiments without hitting hard resource limits (e.g. 4GB memory limit due to using 32bit binaries).

- Clear and well documented programming interface.

- Robustness (simulations should not abort due to programming errors).

- The result values should be presented in a clear way. The values should be presented individually as well as summarized, possibly even with statisticial calculations, like standard deviation. This could be done by an independant tool, which is nevertheless part of the simulation software package.

- The software should be free for academic use.

11.3.2 Examination of Existing Simulation Software

The existing simulation software packages will be examined according to the criteria established in the previous section in order to select the best available software. For each requirement a score between 0 (does not meet requirement at all) to 3 (meets requirement in all possible ways) will be given. The grand total will be an indicator how well the simulation software is suitable for the intended experiments.

NS-2

- **Movement Scenario:** Can be specified in form of a movement trace. Score: 3

- **Traffic Scenario (trace):** Could be programmed, but difficult. Score: 1

- **Traffic generator:** Exists for individual sources. Score: 1

- **Obstacles:** Not implemented. Score: 0

- **Result Values:** Default result is huge tracefile, useful values need to be extracted by custom scripts or other result forms need to be programmed directly into NS-2. Score: 1

- **Clear usage:** Mix of C++ and otcl, not very clear. There is no clearly structured configuration file, but the experiment must be programmed in TCL. Score: 1

- **Well documented:** There is comprehensive documentation but not good enough for the complexity of the software. Score: 2

- **Library RPs:** AODV, DSDV, DSR, TORA. Score: 1

- **Library movement scenarios:** Four simple examples. Score: 1

- **Library traffic traces:** Four simple examples. Score: 1

- **Run-time control:** Not existing. Score: 0

- **Long experiments:** Does not scale well, simulation time far too long and resource limits are reached soon. Score: 1

- **Programming interface:** Exists and is very flexible but it is not clearly structured. It is a mixture from C++ and otcl, but there are no clear interfaces or rules, which parts should be done in which programming language. Score: 1

- **Robustness:** In some scenario combination, NS-2 enters endless loops, which are hard to tell from long simulation times and even harder to debug. This is a major showstopper. Score: 0

- **Clear results:** Default result (tracefile) requires extensive and resource demanding postprocessing. Score: 1

- **Free for use** NS-2 is public domain open source software and thus as free as possible. Score: 3

- **Total score:** 18 (of 48)

NS-2 has several severe drawbacks. NS-2 does not scale very well to scenarios with many nodes and over a long simulation time. Even with very powerful machines, scenarios over 100 nodes and over 3600 seconds simulation time take a very very long time (several days) to complete. Also with larger scenarios resource limits in terms of memory and diskspace can easily exceeded (the tracefile will be about 2 GB per 300 seconds simulation time with 50 nodes, thus a 100 nodes 3600 seconds simulation will already produce a tracefile of about 48 GB. The postprocessing of such a file will also consume a lot of resources and time. It is far better to directly collect the statistical values required for the experiment directly in the simulator. Although it is possible to patch NS-2 accordingly, the unorganized structure makes this very difficult.

However, then main problem with NS-2 is that some modules are unstable in terms of occasionally going into endless loops. Such loops cannot be detected and distinguished from a regular very long simulation run.

Although NS-2 has many advantages and promising ideas, it is not suitable for a large set of simulations to be run in a batch and with clear requirements for result values.

GloMoSim

- **Movement Scenario:** Can be specified in form of a movement trace. Score: 3

- **Traffic Scenario (trace):** Not implemented. Score: 0

- **Traffic generator:** Exists for individual sources. Score: 1

- **Obstacles:** Not implemented. Score: 0

- **Result Values:** Values to be included into the result can be selected. Score: 2

- **Clear usage:** The configuration is clearly structured. Score: 2

- **Well documented:** Only very basic documentation exists. Score:1

- **Library RPs:** AODV, DSR, LAR1, FSR, WRP, ZRP. Score: 1

- **Library movement scenarios:** Two simple examples. Score: 1

- **Library traffic traces:** Exact traces not existing. Score: 0

- **Run-time control:** Not existing. Score: 0

- **Long experiments:** Scales better than NS-2, but hits resource limits more easily (memory), since there is a binary component which is only available as a 32bit binary. Thus the process memory limit of 4GB applies and is hit easily. Score: 1

- **Programming interface:** Is more clearly structured, but not tight enough. Routing protocol modules do not provide a common set of result values, thus making it difficult to compare their performance. Score: 2

- **Robustness:** Aborts simulations due to failed assertions and hitting resource limits. Score: 1

- **Clear results:** Results are presented in a clear list of reasonable size and can be post-processed easily. Score: 2

- **Free for use:** It is free for use to anyone. Most code is open source except the PARSEC core, which is available as 32bit-binary only. Score: 2

- **Total Score:** 20 (of 48)

QualNet

- **Movement Scenario:** Can be specified in form of a movement trace, there are also simple movement generators included into QualNet (implements random waypoint model). Score: 3

- **Traffic Scenario (trace):** Traffic traces can be used for individual sessions (e.g. transmission of an mpeg file). Score: 2

- **Traffic generator:** A set of sample application traffic generators exists, which can be configured to attach to nodes and to generate traffic within some rough constraints. Types available are: CBR, FTP, HTTP, TELNET, VBR, and LOOKUP (models lookup sessions like DNS, LDAP, etc.). Score: 3

- **Obstacles:** QualNet can import terrain data in various formats, which includes elevation and obstacles. The wireless pathloss model takes line of sight availability according to the loaded terrain into account. Movement specified by a tracefile is obviously unaffected, so the external movement generation must take the terrain into account itself. Score: 2

- **Result Values:** Values to be included into the result can be selected from a wide set of statistics available. Score: 2

- **Clear usage:** The configuration file is clearly structured. There is also a GUI tool to help with the configuration. Score: 2

- **Well documented:** Detailed documentation available. Score: 2

- **Library RPs:** AODV,DSR,Fisheye,LANMAR,LAR,OLSR,STAR,ZRP Score: 2

- **Library movement scenarios:** A few very small examples. Score: 1

- **Library traffic traces:** None. Score: 0

- **Run-time control:** Unknown. Score: 0

- **Long experiments:** Scales good, parallelizable. Simulations with 50.000 nodes have been performed. Score: 2

- **Programming interface:** Clear and well documented Score: 2

- **Robustness:** Commercial product with support, even if there are bugs, one can consult the vendor for analysis and fixing. Score: 2

- **Clear results:** Results are presented in a clear way, which can easily be post processed. A set of analysis tools is provided to create diagrams and summarized statistics directly from the result files. Score: 3

- **Free for use:** QualNet is not free, but a commercial product. Even for academic use, license fees apply (although there are discounts available for academic insititutions). Score: 0

- **Total Score:** 28 (of 48)

Simulation Software Selected

Taking the above comparison into account, QualNet is obviously the best choice. However, QualNet is not a free product. Unfortunately, I had no funds available to cover the licensing costs. Thus I could only test QualNet for an evaluation period of 14 days. During these evaluations I even came across a crash, but as the product was not purchased, I could not request support.

Apart from QualNet the next best choice is GloMoSim, which can be considered to be the freely available sibling of QualNet. Also many simulations in previous research have been done using GloMoSim.

The third choice would have been NS-2, but NS-2 suffers from even more severe drawbacks than GloMoSim. While instabilities in GloMoSim are cought by assertions which cause GloMoSim to crash (which is bad enough), I came across instabilities with NS-2, that cause NS-2 to enter endless loops thus keeping NS-2 running for days. They could not easily be identified to be endless loops, I finally identified the problem by attaching a debugger to the running process, but I was not able to track the root condition that caused this loop. Other major disadvantages of NS-2 are the result being only available as a full traffic trace, which requires a huge amount of resources to store and careful post-processing. Alternatively one would have to patch the code. Although some parts of the code are clearly structured, the mixture of tcl and C++ code without clear boundaries and interfaces remains, making it a very complex and difficult task to do it properly. The awkward (but admitted, very flexible) way to configure and use NS-2 by programming the simulation parameters in tcl is another disadvantage.

All in all GloMoSim appeared to be the best choice to perform this small set of sample simulations.

11.4   Scenario Generators Used

Two scenario generators have been used in these experiments, "NoMoGen" (created by Jan Lange[97]), which was already described in section 9.2, and "scengen" [130] a very simple generator which creates movement traces according to the specified mobility model. Their key features are summarized:

11.4.1   NoMoGen

NoMoGen is an event driven scenario generator, that takes the input parameters in form of an XML file. Custom node types can be specified with their movement abilities, also areas within the simulation area can be specified with certain characteristics. Nodes and areas can be attractive or repelling (nodes with other nodes as well as with areas). Areas can be forbidden for nodes, which means nodes cannot enter the area (it can be still attractive, though). NoMoGen not implement a stochastic mobility model. Group mobility can be achieved by setting node attraction parameters. A GUI display of the current progress can be activated optionally.

- No stochastic mobility model.

- Each move of each node is computed according to the attraction factors of its surroundings and an additional random factor (which can be controlled).

- Node type parameters which can be specified are:

    - minimum and maximum speed

    - maximum and minimum (negative) acceleration

    - a random speed variation factor

    - a minimum curve radius factor

    - attraction to other node types

- Area type parameters which can be specified are

    - node attraction factor for individual node types

    - maximum speed for individual node types

    - forbidden for individual node types

- The output is a NS-2 compatible movement trace file

11.4.2   scengen

"scengen" is a lean scenario generator implemented in C++. It takes a scenario specification file (ASCII text) as input and creates NS-2 movement trace data. The scenario is computed according to the selected "motion models". The models can be controlled in their characteristics to some limited extent. For one scenario several sets of nodes can be defined, each set can behave according to a different motion model. The sets of nodes can also be configured to behave as group.

- Stochastic mobility models with group mobility.

- Moves are computed according to the corresponding motion model of the concerned node.

- The following motion models are available:

  - Random Waypoint
  - Fixed Waypoint (single node only, each indiviual waypoint, and thus each move has to be specified)
  - Brownian Motion
  - Pursue Motion (group leader and followers)
  - Column Motion (nodes move always in straight line with varying speed and don't change direction, if they would hit the border, their direction is reversed, this models again only a single node)
  - Gauss-Markov model, using a gauss process to generate velocity change events.

- Typically the following parameters can be set for a set of nodes (not with all models of course):

  - Number of nodes in the set
  - Motion model to follow
  - Area where nodes are placed
  - Acceleration and acceleration time of nodes
  - Minimum and maximum speed of nodes
  - Timeinterval for pause time (for waypoint models)

11.4.3   Traffic Scenario Generation

The traffic pattern was generated using a simple script to provide GloMoSim traffic sources. No external traffic scenario generator or traffic traces from real communications have been used. Instead the GloMoSim provided internal application models were utilized.

FTP, HTTP and TELNET connections have been set up, as described in the following section 11.5.

11.5   Simulation Scenarios Used

The following sections describe the simulation scenarios used and the results of the simulations.

Four scenarios have been examined:

- Town center with cars and people

- Roads on the country side

- A disaster area (as in previous experiments)

- A nature park scenario.

Each scenario was used to create 3 patterns of node placement and movement using NoMoGen. Simulations were then performed with each of these patterns. There was only a single traffic pattern created for each scenario. It was intended to keep the number of degrees of freedom limited in order to avoid a very large set of required simulations and to permit straightforward comparisons.

In addition movement patterns were also created using the *random waypoint* algorithm implemented in "scengen"[130]. The same number dimensions and number of nodes and also the same traffic pattern was used, only the creation of the movement pattern was using the very simplified *random waypoint* algorithm instead of the much more complex generator implemented in NoMoGen.

Thus these simulations can show, if it would actually make a difference to have a more advanced scenario creation.

11.5.1 Town Center with Cars and Pedestrians

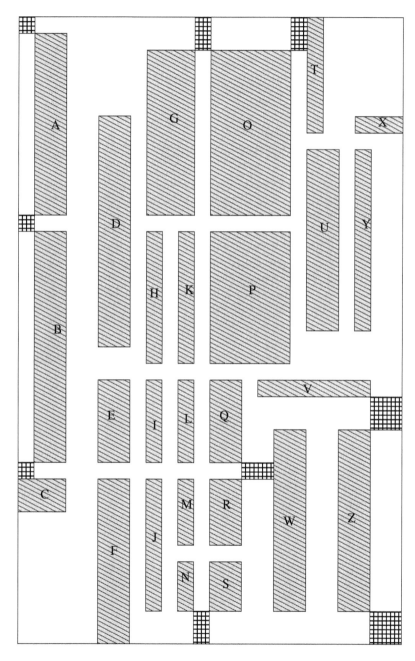

Figure 11.1: A town scenario

Figure 11.1 shows the geographical layout of the town center scenario.

The parameters of this scenario are the following:

Area dimensions:	1200m × 1900m
Number of all nodes:	78
Number of cars:	48
Number of pedestrians:	30
Simulation time:	3600 seconds

The scenario consists of several streets and bigger street areas divided by greenstrips or buildings. Special areas are the two parking grounds and all crossings. For NoMoGen, these areas have certain capabilities like reduced speed and increased chance of a change in the direction. The greenstrip/building areas are prohibited for cars, while pedestrians are repelled from street areas. For the NoMoGen generator the nodes were initially distributed such that the cars started in the parking areas, while the pedestrians were randomly distributed over the whole area.

For the *random waypoint* generator "scengen" the special areas have no effect and all nodes were randomly distributed over the whole area.

The traffic generated for this scenario is created as follows:

FTP connections:	56
TELNET connections:	1016
HTTP connections:	50
HTTP servers:	6

The FTP and TELNET connections are pairwise connections between random nodes. The HTTP connections are between random clients and 6 fixed servers. The 6 servers were chosen randomly but stayed the same over the whole simulation period.

11.5.2 Roads on the Countryside

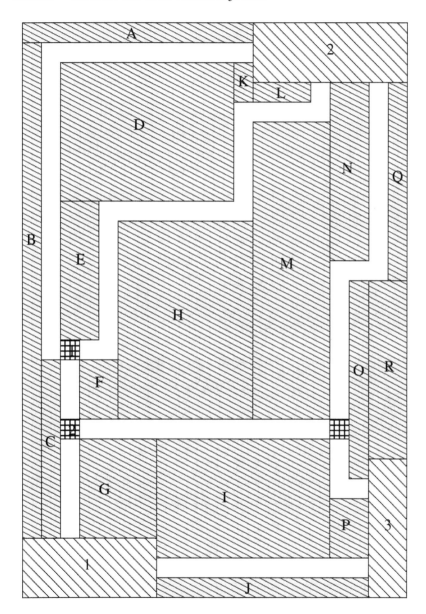

Figure 11.2: A country roads scenario

Figure 11.2 shows the geographical layout of the country roads scenario. It is
similar to the town center scenario, but has a much lower density of streets,
more cars and less pedestrians.

The parameters of this scenario are the following:

Area dimensions:	1000m × 1450m
Number of all nodes:	70
Number of cars:	50
Number of pedestrians:	20
Simulation time:	3600 seconds

The scenario consists of a few roads and some bigger road areas (like town
entrance or big shopping center) and some grass areas in between.

Again, for NoMoGen, these areas have certain capabilities similar to the town
center scenario.

For the *random waypoint* generator "scengen" the special areas have no effect.
The traffic generated for this scenario is created as follows:

FTP connections:	52
TELNET connections:	950
HTTP connections:	51
HTTP servers:	5

The connections are set up in the same way as in the town center scenario.

11.5.3   A Disaster Area

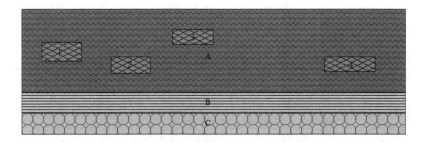

Figure 11.3: A disaster area scenario

Figure 11.3 shows the disaster area. Area A is the main disaster area (site
of a plane crash or a train crash). The small areas within *1 – 4* are the
sites of debris with injured people. These places are the main attractors for
the paramedics, ambulances and choppers. Area B below is a paved area, it
could be a big motorway or a plane runway. Area C is a grassy area. Initially
this scenario had a much longer simulation time, with 5 hours instead of one.

This was chosen to have a scenario also with a more long-term simulation. Unfortunateley it was not possible to finish any simulation of that duration. Thus the simulation time was reverted again to one hour.
The parameters of this scenario are the following:

Area dimensions:	1900m × 600m
Number of all nodes:	157
Number of choppers:	1
Number of ambulance vehicles:	6
Number of paramedics:	150
Simulation time:	3600 seconds

The traffic generated for this scenario is created as follows:

FTP connections:	126
TELNET connections:	1792
HTTP connections:	102
HTTP servers:	12

Again the connections are set up randomly and the HTTP servers stay fixed.

11.5.4 A Nature Park

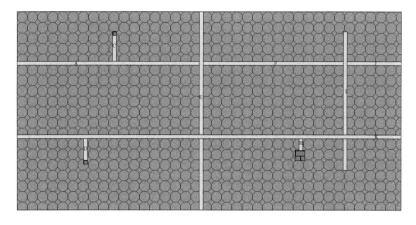

Figure 11.4: A nature park scenario

Figure 11.4 shows the nature park (National Park) or big farm area.
It consists mainly of a very big area of grass, wilderness or crop. There are a few roads cutting through the area and there is one main station and two outposts (ranger stations, or farm headquarters and outpost). The area is of much bigger size than the others and thus the node density is lower. There

are 12 ATVs ("all terrain vehicles", also known as "quads") around the area starting off the headquarters and the outposts. There are also 24 rangers on foot and 50 visitors on foot.

The parameters of this scenario are the following:

Area dimensions:	20000m × 10000m
Number of all nodes:	86
Number of ATVs:	12
Number of rangers:	24
Number of visitors:	50
Simulation time:	3600 seconds

The traffic generated for this scenario is created as follows:

FTP connections:	66
TELNET connections:	1034
HTTP connections:	62
HTTP servers:	6

11.6   Routing Protocols Used for this Simulation

The routing protocols to compare in this simulation were chosen on the basis of availability and stability with GloMoSim. The following routing protocols are implemented with GloMoSim: AODV, DSR, FSR (Fisheye State Routing), LAR1 (scheme 1), WRP and ZRP. Unfortunately ZRP appeared to be unstable with the scenarios used and lead to repeated crashes of GloMoSim, thus only AODV, DSR, FSR, LAR1 and WRP were used in this simulation as provided by GloMoSim 2.03. Since some of these protocols have evolved since their implementation in GloMoSim, the results may vary if a newer implementation is used. However, QualNet would also provide more recent implementations of routing protocols.

The routing protocols are described in more detail in the following sections: AODV: 8.3, DSR: 8.10, FSR: 8.14, LAR: 8.20 and WRP: 8.30.

11.6.1   Representativeness of the Selected Routing Protocols

Although the selected set of routing protocols used for the sample simulations was chosen because these protocols were the ones that are available with the simulation software, it is still of interest, how similar these are to other routing protocols , and if they can be considered to be a representative for a class of routing protocols. If one compares again with the results from section 7.4.3, (figures 7.2, 7.3, 7.4), it can be said, that:

AODV and DSR can be considered to representatives for the class of reactive distance vector protocols.

WRP is a distant member of that class.

FSR may be a representative of the class of link state protocols.

LAR1 is a representative of geographic routing protocols.

Except maybe for AODV, the other protocols are clearly not the best representative for their corresponding class, and may not have been chosen if there would have been alternatives available.

11.6.2 Routing Transport Used in GloMoSim Implementation

To avoid misinterpretation the transport protocol used to transmit routing protocol data for the GloMoSim implementations are described.

AODV The GloMoSim implementation is using raw IP packets to send AODV data (i.e. with AODV being the next layer protocol on top of IP). This is in contrast to more recent specifications of AODV which require AODV data to be sent as UDP packets.

DSR The GloMoSim implementation of DSR works similar to the implementation of AODV, but since DSR specifies the information to be sent along in the IP header this is conform with the DSR specification.

FSR The FSR implementation in GloMoSim is implemented in the application layer using UDP for transport. The FSR specification does not specify the exact layer to implement FSR. It can be done in the network layer or in the application layer.

LAR1 LAR1 is implemented to send routing data by raw IP packets, like AODV and DSR.

WRP WRP is specified to be used on top of the mac layer[114]. The GloMoSim implementation of WRP is similar to FSR and uses UDP for transport.

Only AODV, DSR, FSR, LAR and WRP routing protocols could be simulated, since those are implemented in GloMoSim and actually work. ZRP is claimed to be implemented but all simultions abort with failed assertions.

11.7   Outline of Experiments

The initial plan was to simulate each combination of the sample scenarios with each routing protocol implemented in GloMoSim. The simulation parameters except scenario and routing protocol should remain constant for all simulations. Depending on the simulation time required all simulations should have been repeated a couple of times (10-30) with different random seed and with differently generated scenarios (from the same parameters of course).

Unfortunately this plan failed due to severe flaws in GloMoSim.

Instead only four simulations could have been performed per scenario and routing protocol. Three of them used the more realistic NoMoGen movement scenario generator and varied in the random seed of the generated movement scenario. The fourth used the simpler scengen movement scenario generator. Worse, it was not even possible to run all simulations with a constant random seed, instead every simulation done has most likely used a different one (due to the instability of GloMoSim). As a result the number of samples is so low, that it is not possible to derive a confidence interval for the results, however, the standard deviation could be computed for the three samples of NoMoGen generated scenarios.

Thus the following severe problems come with the simulations:

- Low number of samples

- No constant random seed throughout a batch of simulations

- No true statistical repetitions

The reasons for these (and other) problems are explained in the next section.

11.7.1   Problems with the Simulation Software

The drawbacks discussed with NS-2 are not present in GloMoSim, however GloMoSim has some severe disadvantages, as well.

- The set of implemented Ad Hoc routing protocols in GloMoSim is rather limited.

- GloMoSim itself has programming bugs, too, leading to simulation aborts in some cases. A simulation may fail with one random seed, but succeed with another. Thus, I was forced to script subsequent simulation runs with a varied random seed until a simulation finished successfully. In some instances this required over 1000 attempts until a successful simulation could be run. Some scenarios did never yield a successful simulation so some combinations of scenarios and routing protocols could not be performed at all (like WRP in the "Disaster Area" scenario). Since this problem also lead to the situation that nearly all different simulations were performed with a different random seed, the traffic pattern in these simulations was also different. This is because the exact traffic is generated by GloMoSim itself (thus depending on the used random seed), only guided by the traffic parameters set in the config file. This is a further limitation of the usefulness of the performed simulations.

These bugs have been the main obstacle that severely limited the number of simulations to be done. One possible cause of the crashes (which at least accounted for some) was that GloMoSim (or rather the PARSEC library binary component) is a 32-bit binary, thus limiting the process address space to 4GB, but the scenarios may exceed that limit in some cases.

- The results of the simulations do not match well the desired metrics as described in 2.1.1. GloMoSim only measures values per node, not per packet, thus severely limiting the available statistics. Even worse, different routing protocol modules do measure different kind of values, making it hard to do a direct comparison. There is obviously a big potential for enhancement.

These disadvantages are partially overcome in QualNet, so the experiments described can be repeated even with a broader scope of routing protocols with QualNet, where available.

11.7.2 Actual Number of Simulation Runs

As already described, the actual number of successful simulation runs is rather limited, due to the problems with the simulation tools available. This section summarises the resulting number of simulations:

- For each of the four scenarios, three movement patterns have been created with NoMoGen and one movement pattern has been created with "scengen", resulting in a total of 16 possible scenario-pattern combinations.

- Five different routing protocols have been examined with these 16 scenario-pattern combinations resulting in 80 simulation combinations.

- For each of these 80 simulation combinations at most one single successful simulation run could be performed. There were combinations which failed to yield even a single successful simulation run (e.g. WRP in the Deaster Area scenarios).

- The total amount of successful simulations was 73.

- The total simulation time for all these successful simulations was 565 hours on a Sun Fire V880 with 8 CPUs and 16GB of RAM.

- The average simulation time for successful simulation runs was 7.74 hours, the minimum was 12.6 seconds (random waypoint scenario for the nature park with AODV), the maximum was 150 hours (third NoMoGen generated scenario for disaster area with FSR).

- Due to the many failed simulation attempts (as explained in the previous section) it must be assumed that the "random seed" is a different one for each single simulation.

- The number of attempted simulation runs exceeded 1000 attempts per protocol-scenario-pattern combination in some cases.

- Traffic was created by the internal stochastic traffic generators available with GloMoSim. Since the actual traffic generated also depends on the random seed, it is not equal in all simulations but most likely different.

- The three NoMoGen generated scenarios were considered three samples of the same class of simulations, and average and standard deviation values were calculated among these results. There was no such calculation for the single "scengen" generated scenarios

This very limited amount of samples does not allow to calculate a confidence interval for the results. The best that can be done is to calculate mean and standard deviation for those simulations that have used the same movement generator.

11.8   Results

The result values of the simulations are presented as diagrams in appendix E. This section will examine these results under the aspects described at the beginning of this chapter in section 11.1.

From looking at the results, there is one obvious observation which needs to be taken into account while drawing conclusions. The Nature Park scenario was too challenging for all routing protocols examined. It was nearly impossible for all routing protocols to establish a working link and transmit some data. Thus many measured values (e.g. Mean TCP Delay) need to be interpreted under that assumption. If there was no data transmitted, then there was also no delay, but this is not due to the good performance of the routing protocol.

Again, it has to be kept in mind, that the number of simulations performed (the number of samples) is very low, and thus the results have no strong stochastic significance (cf. section 11.7.2. They can only be considered as rough hints of being in the right or wrong direction.

11.8.1 Do Similar Scenarios Yield Similar Results?

In order to examine if similar scenarios will yield similar results (with the same routing protocol of course), I will first present the similarity results of the scenarios used. The same comparison method has been used as described in chapter 4.

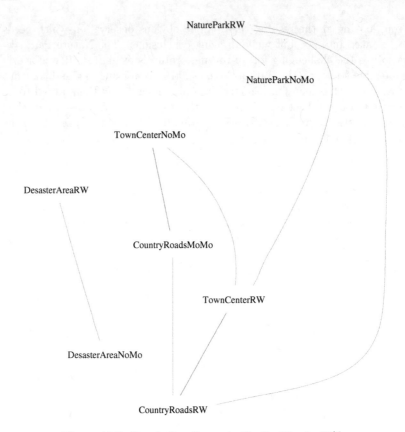

Figure 11.5: Simulation Scenario Similarities $\geq 80\%$

Proto	CRMoMo	CRRW	DANoMo	DARW	NPNoMo
CountryRoadsMoMo	–	0.80	0.78	0.61	0.78
CountryRoadsRW	0.80	–	0.61	0.78	0.72
DisasterAreaNoMo	0.78	0.61	–	0.80	0.72
DisasterAreaRW	0.61	0.78	0.80	–	0.67
NatureParkNoMo	0.73	0.72	0.73	0.67	–
NatureParkRW	0.67	0.84	0.61	0.78	0.86
TownCenterNoMo	0.94	0.77	0.78	0.61	0.80
TownCenterRW	0.77	0.94	0.61	0.78	0.74

Proto	NPRW	TCNoMo	TCRW
CountryRoadsMoMo	0.67	0.94	
CountryRoadsRW	0.84		0.94
DisasterAreaNoMo	0.61		0.61
DisasterAreaRW		0.61	
NatureParkNoMo	0.86		
NatureParkRW	–	0.68	0.85
TownCenterNoMo	0.68	–	0.80
TownCenterRW	0.85	0.80	–

Of course the NoMoGen and RandomWaypoint generated scenarios are found rather similar, because their characteristic parameter were chosen the same, just their movement patterns differ. However, for this section, we consider them separate and thus we can conclude that the Town Center and the Country Roads scenarios are very similar, while the Disaster Area scenario is different. The Nature Park scenario is still a bit similar to the Country Roads and Town Center scenarios (at least the RandomWaypoint part).

Taking a look at the results of the scenarios in appendix E, and comparing the left diagrams with each other (the left diagrams are the results of the NoMoGen generated scenarios, while the right column shows the scengen/Random Waypoint generated scenario result diagrams), we can observe the following:

- For the following measured values, the similar scenarios, Town Centre and Country Roads yield the same relative values for the examined routing protocols (i.e. the order of routing protocols from the best to the worst remains the same), and even the absolute values are not far off, while this is not clearly the case for the other two scenarios (for NoMoGen as well for scengen generated scenarios): **Routing Overhead**, **Route Setup Ratio**, **IP Delivery Ratio**, **MAC Broadcast Ratio**, **TCP Packets Received**, **TCP Packets Retried** (although there is a slight change of order between FSR and DSR in the NoMoGen scenarios), **Broken Links** (but Disaster Area has also similar results), **FTP Throughput** and **HTTP Throughput**.

- **Mean TCP Delay** does not show such a clear similarity between Town Centre and Country Roads, but the standard deviation for the NoMoGen scenarios is much higher, possibly more than three sample simulations might have led to a clearer result.

- **TELNET Throughput** shows no significant similarity between Town Centre and Country Roads.

Conclusion   This result confirms that similar scenarios will yield similar results for almost all measured performance values. Note, that only similar scenarios with the same scenario generator have been compared. This is a confirmation, that the similarity function to compare scenarios is actually useful and in the right direction.

11.8.2 Differences Between Realistic and Simple Scenarios

Even though in the previous section it was compared if similar scenarios yield similar results, only scenarios generated with the same movement generator have been compared to each other, an even more important question is, if the choice of a movement generator has a significant influence on the results, especially if a very simple mobility model (like Random Waypoint implemented in *scengen* [130]) is compared to a complex one, as implemented in *NoMoGen*[97]. As described earlier, one of the aims of this simulation is to find at least some sample evidence, that it makes a difference if realistic scenarios (as described in chapter 4) are used instead of simple scenarios (like random waypoint, cf. 4.2.2).

In appendix E, the corresponding results are presented next to each other in the same line (NoMoGen on the left, scengen on the right).

The following can be observed:

- In the following measured values, the order of routing protocols (from the best performing to the worst) changes:

  Routing Overhead: Mainly DSR performs different using either a complex or a simple mobility model.

  Route Setup Ratio: Differences for Town Center and Country Roads, no differences for Deaster Area and Nature Park.

  Mean TCP Delay: Differences in the Town Center, Country Roads and Disaster Area (although less distinct and with an increased standard deviation on the NoMoGen scenarios for Country Roads and Disaster Area).

  TCP Packets Received: Big differences for Town Centre and Country Roads, slight differences on Disaster Area.

  TCP Packets Retried: Little differences in the order for Town Centre, bigger differences for Country Roads.

  FTP Throughput: Again big differences for Town Centre and Country Roads. Also big differences for Nature Park. Actually some FTP connections could be established in the Random Waypoint scenario, while none could be established in the NoMoGen generated scenarios.

  HTTP Throughput: Big difference for Town Centre and some small differences for Country Roads. Also for Nature Park, the WRP protocol achieved to transmit some data via HTTP in one of the

NoMoGen generated scenarios, but none in the scengen generated scenario.

TELNET Throughput: Slight differences for the Town Centre scenario and Country Roads scenario, interestingly in the Disaster Area scenario the differences are bigger this time. However, the TELNET traffic was only a minor fraction of the overall traffic of the scenarios.

- In the following measured values, the absolute result values differ by about one order of magnitude (or more):

Route Setup Ratio: AODV, DSR and LAR1 in Town Center and Country Roads.

TCP Packets Retried: For Town Centre and Country Roads an absolute difference of close to an order of magnitude is reached.

Broken Links: For Town Centrea and Country Roads, also the number of broken links differs about an order of magnitute. This corresponds with the number of TCP packets retried.

FTP Throughput: In the Town Centre scenario, the AODV throughput differs nearly about an order of magnitude. Since there was no throughput at all for the NoMoGen generated Nature Park scenario, but a significant throughput was achieved in the scengen generated scenario.

TELNET Throughput: Again big differences for the Nature Park scenario, some telnet connections could be established in some of the NoMoGen generated scenarios by AODV and LAR1, while none could be established in the scengen generated scenario.

- For the following measured values, the results are similar and there are no significant differences:

Route Setup Ratio: For the Disaster Area and the Nature Park scenario.

IP Delivery Ratio: For all scenarios.

MAC Broadcast Ratio: For all scenarios.

TCP Packets Retried: Only for the Disaster Area scenario, there is no significant difference in TCP packets retried.

Broken Links: Corresponding to the TCP packets retried, the number of broken links for the Disaster Area scenario shows no significant difference.

HTTP Throughput: Again only in the Disaster Area scenario the HTTP throughput appears similar between NoMoGen and scengen scenarios, although the standard deviation is again very high in the NoMoGen generated scenarios.

Conclusion Although there are some measured values that appear to be unaffected by the generated movement scenario (like IP Delivery Ratio and MAC broadcast ratio), and although the Disaster Area scenario in particular seems less influenced by the movement scenario[2], in general most measured values for most routing protocols and scenarios show a significant difference, either in absolute values or relatively (in the order of best performing routing protocols).

Thus it appears to be justified to ask for more complex movement scenarios for a proper evaluation of routing protocols for Mobile Ad Hoc Networks.

11.8.3   Do Similar Routing Protocols Yield Similar Results?

In order to make use of the similarity relation established between routing protocols, it needs to be examined, if similar routing protocols (in that terms) will also show similar performance results.

The following diagrams display the similarity of the routing protocols used in the simulation. It is an excerpt from chapter 7.

From the results from chapter 7, as displayed in the diagram, it will follow, that most similar behaviour can be expected from AODV and DSR, and AODV and LAR, while other combinations may not necessarily show similar behaviour, with the least similar behaviour between FSR and DSR and FSR and LAR.

---

[2]This is no suprise, since the Disaster Area scenario basically allows the nodes to roam freely over the whole area, just placing some attractors and influence the capabilities on different ground. Thus, there are far less restrictions than in a town or road scenario compared to the random waypoint model (which ignores these restrictions).

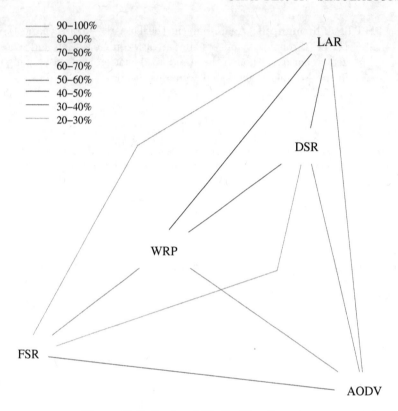

Figure 11.6: Protocol Similarities Overview

Proto	AODV	DSR	FSR	LAR	WRP
AODV	–	0.91	0.38	0.67	0.64
DSR	0.91	–	0.28	0.58	0.55
FSR	0.38	0.28	–	0.29	0.62
LAR	0.67	0.58	0.29	–	0.42
WRP	0.64	0.55	0.62	0.42	–

Examining again the results in appendix E under this aspect, it can be observed, that AODV and DSR do not yield a clear similar result in most of the measured result values, except IP Delivery Ratio. Instead in some result values, AODV and LAR show a similar result, e.g.: TCP Packets Received, also a bit Mean TCP Delay, FTP Throughput

However, in some result values, DSR and FSR show significant differences: Routing Overhead, and IP Delivery Ratio. Also FSR and LAR show differences in some result values, as well: IP Delivery Ratio, Mean TCP Delay (although only small differences), TCP Packets Received and TCP Packets retried (again only small differences).

Conclusion   Unfortunately there is no clear result, if routing protocols which are considered similar, do actually show a significant similar behaviour, neither do protocols which are considered less related to behave much differently. There are several consequences from this situation:

- As with most results from these simulations, they need to be repeated with more samples to get a better statistical ground and more significant results. These results may still be subject to random side effects.

- The similarity relation for routing protocols may need to be improved.

11.8.4   Performance of Examined Routing Protocols

Although some aspects of the results have already been carefully considered, one big question remains, which protocols did perform best in which scenario according to the performed simulations.

In order to get a useful result, each scenario will be examined separately. For each performance measure used, each routing protocols will receive a score, according to its performance. The scores will be very simple, and correspond to the rank. Thus, the best routing protocol, will achieve a score of 4, while the worst will get 0. Thus for most scenarios, the total score over all routing protocols is 100, except for the Nature Park scenarios, because not all scores could be distributed, but some protocols received an equal score (mostly 0), due to similar (bad) performance.

The Broken Links value will be excluded from this evaluation, because it is only available for two of the five routing protocols. Route Setup Ratio is also omitted, since it is a rather informative result, but no real performance criterion and it is not available for all routing protocols anyway. Because of the problems with IP Delivery Ratio (cf. section 11.2) it is not used, as well.

In the Nature Park scenario, in many cases no connections could be established at all. In this case all concerned routing protocols will receive a score of 0.

Finally two measured values, Mean TCP Delay and FTP Throughput, are considered most imporant and thus their score will be doubled.

Town Centre - NoMoGen

Value	AODV	DSR	FSR	LAR1	WRP
Routing Overhead:	1	0	3	4	2
MAC Broadcast Ratio:	3	0	1	4	2
Mean TCP Delay ($\times 2$):	0	6	4	2	8
TCP Packets Received:	4	0	3	1	2
TCP Packets Retried:	1	2	3	0	4
FTP Throughput ($\times 2$):	2	6	4	0	8
HTTP Throughput:	0	4	3	1	2
TELNET Throughput:	2	4	1	0	3
Total:	13	22	22	12	31

Thus yielding the following ranking:

Rank	Protocol	Score
1.	WRP	31
2.	DSR and FSR	22
3.	AODV	13
4.	LAR1	12

Town Centre - scengen

Value	AODV	DSR	FSR	LAR1	WRP
Routing Overhead:	1	3	2	4	0
MAC Broadcast Ratio:	3	2	1	4	0
Mean TCP Delay ($\times 2$):	0	6	2	4	8
TCP Packets Received:	4	2	1	3	0
TCP Packets Retried:	0	2	3	1	4
FTP Throughput ($\times 2$):	0	4	6	2	8
HTTP Throughput:	0	3	1	2	4
TELNET Throughput:	1	3	4	0	2
Total:	9	25	20	20	26

Thus yielding the following ranking:

Rank	Protocol	Score
1.	WRP	26
2.	DSR	25
3.	FSR and LAR1	20
4.	AODV	9

Country Roads - NoMoGen

Value	AODV	DSR	FSR	LAR1	WRP
Routing Overhead:	2	0	3	4	1
MAC Broadcast Ratio:	3	2	1	4	0
Mean TCP Delay ($\times 2$):	6	2	4	0	8
TCP Packets Received:	4	2	3	0	1
TCP Packets Retried:	1	3	2	0	4
FTP Throughput ($\times 2$):	2	6	4	0	8
HTTP Throughput:	1	3	2	0	4
TELNET Throughput:	1	2	3	0	4
Total:	20	20	22	8	30

Thus yielding the following ranking:

Rank	Protocol	Score
1.	WRP	30
2.	FSR	22
3.	AODV and DSR	20
4.	LAR	8

Country Roads - scengen

Value	AODV	DSR	FSR	LAR1	WRP
Routing Overhead:	1	3	2	4	0
MAC Broadcast Ratio:	3	2	0	4	1
Mean TCP Delay ($\times 2$):	0	6	4	2	8
TCP Packets Received:	4	3	1	2	0
TCP Packets Retried:	0	2	3	1	4
FTP Throughput ($\times 2$):	0	4	6	2	8
HTTP Throughput:	0	4	3	2	1
TELNET Throughput:	4	0	3	1	2
Total:	12	24	22	18	24

Thus yielding the following ranking:

Rank	Protocol	Score
1.	DSR and WRP	24
2.	FSR	22
3.	LAR	18
4.	AODV	12

Disaster Area - NoMoGen

Since no simulations with WRP could be performed in the Disaster Area scenario, it has been excluded from the results. Thus the scores will range from 1 to 4 instead of 0 to 4.

Value	AODV	DSR	FSR	LAR1
Routing Overhead:	1	2	3	4
MAC Broadcast Ratio:	2	4	1	3
Mean TCP Delay ($\times 2$):	8	4	6	2
TCP Packets Received:	4	2	1	3
TCP Packets Retried:	3	4	1	2
FTP Throughput ($\times 2$):	6	8	2	4
HTTP Throughput:	3	2	4	1
TELNET Throughput:	4	3	2	1
Total:	31	29	20	20

Thus yielding the following ranking:

Rank	Protocol	Score
1.	AODV	31
2.	DSR	29
3.	FSR and LAR	20

Disaster Area - scengen

Value	AODV	DSR	FSR	LAR1
Routing Overhead:	1	3	2	4
MAC Broadcast Ratio:	2	4	1	3
Mean TCP Delay ($\times 2$):	4	6	8	2
TCP Packets Received:	4	1	2	3
TCP Packets Retried:	2	4	1	3
FTP Throughput ($\times 2$):	6	8	4	2
HTTP Throughput:	2	4	1	3
TELNET Throughput:	3	2	4	1
Total:	24	32	23	21

Thus yielding the following ranking:

Rank	Protocol	Score
1.	DSR	32
2.	AODV	24
3.	FSR	23
4.	LAR	21

Nature Park - NoMoGen

Value	AODV	DSR	FSR	LAR1	WRP
Routing Overhead:	3	2	0	4	1
MAC Broadcast Ratio:	1	2	3	0	4
Mean TCP Delay ($\times 2$):	0	0	0	0	8
TCP Packets Received:	3	0	2	4	1
TCP Packets Retried:	2	0	3	1	2
FTP Throughput ($\times 2$):	0	0	0	0	0
HTTP Throughput:	0	0	0	0	4
TELNET Throughput:	4	0	1	3	2
Total:	13	4	9	12	22

Thus yielding the following ranking:

Rank	Protocol	Score
1.	WRP	22
2.	AODV	13
3.	LAR	12
4.	FSR	9
5.	DSR	4

Nature Park - scengen

Value	AODV	DSR	FSR	LAR1	WRP
Routing Overhead:	3	1	0	4	2
MAC Broadcast Ratio:	0	2	3	1	4
Mean TCP Delay ($\times 2$):	0	0	0	0	0
TCP Packets Received:	3	0	2	4	1
TCP Packets Retried:	3	0	3	1	4
FTP Throughput ($\times 2$):	4	0	6	2	8
HTTP Throughput:	0	0	0	0	0
TELNET Throughput:	0	0	0	0	4
Total:	13	3	8	12	23

Thus yielding the following ranking:

Rank	Protocol	Score
1.	WRP	23
2.	AODV	13
3.	LAR	12
4.	FSR	8
5.	DSR	4

Score Summary

The following table summarizes the achieved scores and also presents a total score for each routing protocol:

Value	AODV	DSR	FSR	LAR1	WRP
Town Centre NoMoGen	13	22	22	12	31
Town Centre scengen	6	25	20	20	26
Country Roads NoMoGen	20	20	22	8	30
Country Roads scengen	12	24	22	18	24
Disaster Area NoNoGen	31	29	20	20	− (20)
Disaster Area scengen	24	32	23	32	− (20)
Nature Park NoMoGen	13	4	9	12	22
Nature Park scengen	13	3	8	12	23
Total:	132	159	146	143	156 (196)

Looking at these plain scores, DSR would be the winner. However, WRP comes extremely close behind and considering that WRP could not be evaluated in the Disaster Area scenario, WRP can be declared the best performing routing protocol (which becomes particulary obvious if one would assume WRP would have received the average score of 20 in the Disaster Area scenarios).

This is actually quite unexpected. WRP is one of the oldest routing protocols designed and seemingly more sophisticated algorithms have been developed with the intention of being able to cope better with the challenges in a MANET. AODV and DSR have received a lot of attention in the scientific world and have been developed much further. It has to be assumed that the version implemented in GloMoSim is a very early version of DSR and AODV, not corresponding to the latest internet drafts or experimental RFCs of these protocols. So an up to date implementation of these protocols may yield a different result [3]. Still, even those early versions have been developed after

---

[3]WRP has also been developed further leading to WRP-lite, BEST and DST, cf. section 8.30

WRP and so it is quite a surprise, that WRP performed so well compared to the other protocols.

The advantage of WRP does not even seem to depend on the scenario (even though WRP could not be tested in the Disaster Area scenario), because even in such different scenarios like Town Centre and Nature Park, WRP performed superior.

DSR is close behind in many scenarios, except the Nature Park scenarios, it is in the top group and can also be considered a sound alround protocol, useful in many scenarios.

FSR and LAR1 do not perform very good, but mediocre, with FSR being ahead of LAR1 in most cases. Both show their weaknesses in the particulary challenging Nature Park scenarios (but the other routing protocols did have the same problems).

The big surpise is the exceptional low rank of AODV, which is considered to be one of the more advanced routing protocols. One reason is that AODV generates a lot of routing overhead and sometimes long delays. Of course the method to distribute scores by rank, may not be ideal, since AODV lost scores but only by a very small difference in performance to the next competitor. Obviously this comparison could be done a lot more fine grained, but this is only worthwhile, if simulation result data is more reliable than those available.

More recent implementations of the examined routing protocols, especially AODV, may yield even more interesting results.

11.8.5 Summary of Results

The results (subject to their statistical significancy) did clearly confirm that the methodology to use a similarity relation is a useful way to limit the number of simulation scenarios and thus the number of simulations to be performed.

The results did also confirm my doubts that very simple scenario models (like random waypoint) are actually useful and allow conclusions about more realistic and real-world applications. In fact, such simple scenarios may not be useful at all.

The similarity relation method did not clearly succeed to limit the number of routing protocols. It could not shown, that similar routing protocols yield similar results. However, the contrary could not be shown either. Thus, the method may still be useful in general, but the particular similarity relation may need to be improved.

Chapter 12

Summary and Conclusion

12.1 Summary

In this book I have analyzed many past evaluation studies about MANET routing protocols. The problems with ad hoc networking and with simulation based evaluation have been described. Especially the mobility models used in evaluation studies have been examined and some problems have been underlined. The fact that any evaluation must take place in context of an intended application has been emphasized.

A method to compare evaluations and applications and to match evaluations with applications has been developed and implemented. Criteria to compare and evaluate and match simulation scenarios, applications and MANET routing protocols have been established. Comparisons and matching have been performed and the results have been presented. Other aspects of simulation based evaluation have been examined and discussed.

A comprehensive amount of MANET routing protocols has been studied, characterized and compared. I have suggested guidelines to further improve the quality and applicability of simulation based evaluation of MANET routing protocols by proposing an enhanced node-interactive mobility model, which has been implemented in a student project. Finally, I have investigated general problems with simulations and another student project was issued to overcome these problems with the lack of automated simulation performance.

A small set of sample simulations has been performed and the results examined to support or contradict the assumptions made earlier. The results confirmed that the similarity relation of simulation scenarios allows to draw conclusions about the relative performance of MANET routing protocols. They further confirmed that the mobility model and thus the movement sce-

nario used in a simulation scenario does indeed affect the simulation results
and consequently more complex and realistic mobility models should be used
in future simulations.

However, the results could not clearly tell if the similarity relation between
routing protocols defined will allow to draw conclusions about their performance. It is very well possible, that a relation, that would allow such
conclusions cannot be defined from the external characteristics of a routing
protocol. Further, the sample simulations provided practical insights in the
numerous problems and difficulties of simulation based evaluation of routing
protocols for Mobile Ad Hoc Networks.

12.2 Conclusion

This work has shown, that any evaluation needs to take an application context into account. This is especially true with evaluating MANET routing
protocols, since the possible applications cover a broad range and differ a lot
in their key characteristics. As long as there is no analytical way to evaluate
a routing protocol in an application context (and it is doubtful, if there will
ever be such a way, that can be used in general), simulation based evaluation
is the only way. In order to get reasonable results, steps must be taken to
ensure that the simulation models the intended behavior well enough. This
can be achieved by using mobility models that match realistic movement of
the used node types better (e.g. using obstacles, respecting acceleration and
individual movement strategies) as shown but also using traffic models that
cause realistic communication patterns between the nodes.

Further the currently awkward and time consuming way of setting up simulations needs to be simplified and automated as far as possible, thus allowing
more simulations to be performed in single studies using better models. Of
equal importance is a robust and performant simulation tool, that yields consistent statistical result values for all implemented routing protocols (which
was not the case with GloMoSim) and allows to simulate reproducible instances.

12.3 Outlook

Although this research work did perform some simulations, there are still
many starting points for more studies to be done now in a better and more
efficient way. The tools presented can be improved (e.g. in other student
projects) to be of good value for the research public. More precise application

scenarios may be defined and simulation scenarios should be modeled after these.

Such a scenario specification can be used as an input to the scripting framework, which in turn uses a scenario generator that takes the results of this work into account.

A comprehensive study evaluating more protocols (using up to date implementations) for a set of given applications with much more statistical samples (i.e. more actual simulation runs) can be performed to show clearly, which algorithms offer best performance under realistic conditions.

Especially the results of the performance comparison of sample routing protocols with WRP outperforming AODV suggest further investigation.

Another interesting question for further investigation would be to find out, which are the best possible performance results achievable for a routing protocol given a specific scenario (i.e. the ideal case) and how close could real MANET routing protocol implementations get.

Such a study can also help to fine tune the similarity relations and to get more evident information, if a useful similarity relation between routing protocols can be established, in order to reduce the number of future simulations.

After all, the comparison and evaluation method using similarity may be applied to other problems with similar properties, i.e. multidimensional comparison and evaluation problems with many elements that benefit from a significant reduction of the search space, especially if individual experiments are comparatively expensive. Examples could be in molecular genetics, chemical engineering or pharmaceutical research.

I hope this book proves to be a useful contribution to the research community studying Mobile Ad Hoc Networking.

Chapter 13

Acknowledgments

I would like to thank the following people. Without them, my dissertation and consequently this book would not have been possible. My supervisor Prof. Dr. E. Jessen for great support and continuing confidence. Dr. M. Jobmann for counsel and critical comments. My colleagues V. Nutzinger, H.O.Riethmayer, J. Lesny, M. Herrmann and A. Paul for taking my regular work off my back and their valuable opinions and input. My wife Sabine and my parents for continued motivation to finish this work.

Appendix A

Definitions

A.1 Terms Generally Used in this Book

This sections defines and clarifies some of the more frequently used terms to avoid any ambiguities and misunderstandings.

Attribute The term *attribute* refers to an individual attribute of a certain →*characteristic* of either a →*simulation scenario*, an →*application scenario* or a routing protocol.

Characteristic Commonly used to describe a particular, measurable *characteristic* of either a →*simulation scenario*, an →*application scenario* or a routing protocol. A characteristic can consist of one or more →*attributes*.

Evaluation Used in the context of a →*simulation scenario* based →*experiment*, or for a benchmark comparison of simulation scenarios with the aim to find out if some of the examined items (scenarios, routing protocols) are performing in a better way, than the others.

Experiment The term experiment is used to describe the whole process of creating a →*simulation scenario*, actually performing the simulations and deriving results. It is also used for real world (i.e. not simulated) experiments, e.g. the CMU testbed[111].

A.2   Terms Used in Simulation and Application Scenario Context

This section will define commonly used terms and phrases which are mainly used in the context of scenarios, that are used in simulations to model real applications and in real experiments.

Application Scenario: An *application scenario* describes the characteristics of an intended application for a mobile ad hoc network including it's →*movement* and →*traffic scenario*. Sample application scenarios are described in chapter 5.

Area, Observed Area: The observed area of the evaluation (simulation, testbed or real installation). The shape of the *area* is usually a rectangle, but doesn't necessarily need to be.

Border: The border of the →*observed area*.

Border Behavior: The behavior of →*nodes*, if they approach the →*border*. This is an important characteristic of a →*mobility model*. Possible border behavior is described more detailed in section 4.5.5.

Group: A *group* is a set of *nodes*, which share some common characteristics and which may move together in some way. Usually a group has a common *movement strategy*, but does not need to. The actual characteristics may vary and are defined by the group specification.

Group Mobility: *Group mobility* or a *group mobility model* is part the general *mobility model*. It is defined as the set of rules that allow nodes to form and maintain *groups* and may influence the movement decisions of group members.

Mobility Metric: A metric to measure the *degree of mobility*. Examples for a simple mobility metric could be: average speed of the nodes or average →*pause time*. Much more complex mobility metrics are possible and discussed in sections 4.2.3 and 4.5.2.

Mobility Model: The *mobility model* is defined as the set of rules, that determine the movement of the →*nodes*.

Movement Strategy: The *movement strategy* is an important part of the *mobility model*. I define the *movement strategy* as the set of rules, that determine the intended destination of each move of each node, and also the intended movement speed. The actual movement speed and direction will also be influenced by other parameters (like maximum speed in a certain *region*), which are not part of the *movement strategy*.

Movement Scenario: The *movement scenario* consists of the union of the *observed area* including *subareas* and the *mobility model* plus some additional characteristics as described in section 4.1. It is not a *specific* scenario, which would be one single strictly determined way, how nodes behave[1]

Node: A node is a MANET enabled device, attached to an object that can move and act individually. Examples for *nodes* are:

- A person, that carries a cellular phone, a notebook computer or an organizer with MANET capable communications hardware.

- Such persons, but using a bicycle or public transport.

- A car fitted with MANET capable communications hardware.

- A tank or other military vehicle (possibly unmanned) fitted with MANET capable communications hardware.

Also aircraft, helicopters and ships could generally be regarded as *nodes*, but this paper does not take these types of nodes into account.

Node Density: The node density in the observed area, as commonly defined:

$$\frac{n}{A}$$

with n being the number of nodes, and A being the amount of space, covering the observed area which contains all nodes n.

Pause Time: This term was used first in the introduction of the first *random waypoint* mobility model in [86]. The *pause time* is a fixed time, which a node is waiting after each move. The *random waypoint* is described more detailed in section 4.2.2.

Performance: Since the goal of simulations and evaluations is to determine, which routing strategy *performs* best, under which conditions, the term *performance* is used very often. Since there is not a single performance criterion, I define performance in the sense of an overall performance which takes the following measures into account, which

---

[1]A specific scenario would be the kind of scenario that is ultimately used in a single simulation.

are commonly used throughout the evaluations: **end-to-end delay of a packet**, **average and maximum throughput and goodput**, **initial connection setup latency**, **routing overhead**, **path length** and **overhead due to suboptimal paths**. There are different uses of these values in the various studies, like **throughput** and **overhead** can be measured in terms of *Bytes* or *Packets*. For my use of *performance*, I do not prefer one or the other possible definition, since that is up to the evaluations itself. With *performance*, I mean the performance of a routing protocol qualitative under the various aspects mentioned here. See also section 2.1.1 for individual performance metrics.

Region or Subarea: This is a usually smaller (but not larger) part of the →*observed area*. It may have special properties that affect node movement and communications.

Scenario: There are →*movement scenarios*, →*traffic scenarios*, →*simulation scenarios* and →*application scenarios*.

Simulation Scenario: A *simulation scenario* describes the characteristics of a simulation experiment including it's →*movement* and →*traffic scenario*. Simulation scenarios are examined in chapter 4.

Traffic Scenario: A *traffic scenario* consists of a specification of the traffic generated by each node to each other node, including type, required bandwidth, duration and exact transmission times. The units can be in packets or bytes. Depending on the intent, the actual traffic transmitted may be specified alternatively the payload, that each node wants to transmit may be specified.

A.3   Terms Used in Routing Protocol Context

This section describes terms and phrases, that are used in descriptions and definitions of routing protocols for mobile ad hoc networks.

Active Receiver: A node that is receiving data at the moment or is part of a session and likely to receive data in the near future. The term *active receiver* is only used in ADV (cf. Section 8.2).

Anchor: A certain geographical position in the network's area. Anchors are used for geographical routing with *Terminodes* (cf. Section 8.27).

Beacon, HELLO message: A (usually) periodic →*local broadcast* message emitted from a node, destined for it's →*neighbors* to announce itself in the neighborhood. In some routing protocols, such a beacon may carry additional information.

Bordercast: A term from the Zone Routing Protocol ZRP (cf. Section 8.31). A message is transmitted to one or more nodes on the border of the current routing zone, where it might be transmitted further.

Broadcast, local: A local broadcast is a broadcast message, that can be received from any node within direct reach of the sender. It is not intended to be retransmitted by the receivers.

Broadcast, limited: A limited broadcast can be retransmitted/forwarded, but only to a subset of nodes in the network, limited by the maximum number of hops (time to live) of the packet. Some routing protocols send limited broadcasts into a certain direction.

Broadcast, net-wide: This broadcast ist retransmitted, until every node in the whole network has received the message at least once. This type of broadcast causes a very high network load and may influence the transmission of other messages.

Cluster: A group of nodes, that act together in some way. Usually a cluster is represented by a single node, the *Clusterhead*. Clustering is used in clustered and hierarchical routing algorithms. Clustering allows to form structures even in a very dynamic environment.

Clusterhead: The representative node of a cluster. On a higher routing level, routing happens between the clusterheads. On the next higher level again clusters will be built but out of the clusterheads of the level below.

Distance Effect: The *distance effect* is that two nodes appear to move slower with respect to each other if they are more distant. DREAM (and certain other protocols) make use of that fact. Routing or position information for distant node does not need to be as accurate as for nodes in close distance.

Distance Vector Routing: Simple, table based routing. Each destination is entered into the routing table with the next hop and a distance metric. The topology of the network is unknown.

Expected Zone: The geographical region in the network, where a certain node is expected to be. The expected zone is predicted from the last known movement characteristics of a node. The expected zone is used to derive a →*request zone*, needed for location based protocols.

Flooding: A message is *flooded* through the whole network. This is another term for a *net-wide broadcast*.

Friend Node: In *Terminode routing* a node, which is in close vicinity, but does not need to be in direct communication range. A friend helps determining an *anchored path* in AGPF (cf. section 8.27).

Gateway: A node within a *Cluster*, often part of more than one clusters, which route messages from one cluster to another.

GPS: The *geographic position system*, a satellite based position service operated by the US military. It enables a GPS receiver to determine its position.

Group Mobility: Nodes can form groups and move together as a group. This is a likely event in realistic scenarios and needs to be modeled in the scenario model. Group mobility has significant impact on routing performance, depending on the protocol and it's ability to handle group mobility. See also the →*group mobility* entry in section A.2.

IEEE 802.11: IEEE standard family for wireless LAN communication. It defines the *distributed coordinate function* (DCF) or the *point coordinate function* (PCF) as channel allocation method. PCF can not be used in ad hoc networks, since it would require a central instance (like an Access Point), but DCF is very common. DCF defines a RTS/CTS (request to send/clear to send) handshake to allocate a channel, thus circumventing the hidden terminal problem.

ILS: An *idealized link state* algorithm. Such an algorithm is mentioned in several papers as a reference for comparison, but is never specified in detail.

Link Layer Notification: A mechanism, that allows the routing module to be notified of local link breaks or new links (a node moves out of reach, or a node moved into reach) from the link layer.

Location Dependent Address: An address, which depends on the geographical location of a node. It must be determined by a *location* *service.* Location based routing protocols, like Terminode/AGPF (cf. Section 8.27) make use of LDAs.

Location Service: See →*Position Service.*

MANET: Abbreviation for *Mobile Ad Hoc Network*, also name of the corresponding IETF working group.

Maximum Hop Count: The maximum amount of hops a packet is allowed to travel. Often also called *time to live.*

Multi-point Relay: A dedicated node, that relays traffic for other nodes in OLSR (cf. Section 8.23).

Neighbor: Any other node participating in the MANET within direct communication range of a particular node.

Node: A device, capable of communication over a wireless link and attached to some (in most cases mobile) unit, like a person or a car, participating in the MANET.

Overhead: In general, the amount of data transmitted in addition to the payload. For a more thorough discussion, see section 2.1.1.

to overhear: A node can *overhear* messages not destined for it, by setting it's interface into *promiscuous* mode. A node can benefit from routing information for other nodes. Additional routes may be learned and routes may be updated before the routes are needed and fewer route discovery processes may be needed. DSR makes use of this feature.

Parent Node: The node's current uplink in a route. See also →*Precursor.*

Partial Topology: Several link state routing protocols do not maintain full topology information but only *partial topology*, sufficient for efficient routing, while full topology would only use much more resources.

Position Service: A service, that can provide positional information for nodes in a mobile network. The position service needs to provide the position of any node to any node (GPS only provides the position for the node itself). Depending on how the position data is obtained, stored and distributed, there are *all-for-all*, *some-for-all* and *some-for-some* types of position services [113].

Precursor: The precursor node of the current node in a route to a destination. If a route becomes invalid, the precursors of the current node
may need to be notified to update their routing entries. AODV (cf.
Section 8.3) explicitly uses a *precursor list*.

Proactive: Routing protocols are considered as *proactive* if they constantly
maintain routing information for all routes, regardless, if in use or not.
Maintenance can be event-driven (also named *table driven*) or in regular
intervals.

Reactive: Routing protocols, which obtain and maintain only routes that
are currently needed, are called *reactive* or *on demand*. They can cache
learned routes, but if a route is unknown, a route discovery process
needs to be initiated.

Request Zone: A geographical zone in the network, that covers all the
nodes a *route request* should be sent to. The request zone is used in
geographical routing algorithms, which results in a *limited broadcast*
(cf. →*expected zone*).

Rooftop Network: An immobile ad hoc network. Nodes are deployed on
rooftops, but do not move once deployed. They still need to organize
themselves in an ad hoc fashion. (This study does not consider rooftop
networks.)

Route Cache: A local cache in a node used by *reactive* routing protocols,
to cache discovered routes. The routes will eventually time out from
the cache, or be expunged if the route is detected to be invalid.

Routing Loop: The route forms a loop, such that packets are routed in
the loop and possibly never reach their destination if the loop persists.
Routing loops need to be avoided for successful routing. The common
way to avoid the formation of routing loops is the use of sequence
numbers in routing update packets.

Route Request: Important part of a route discovery process. A route request is usually a →*limited broadcast* message destined for the target
node of the required route. If the destination receives the route request,
it will answer with a route reply. The route request messages (also
named *broadcast query*) are abbreviated as RREQ, REQ or QRY in
the most common routing protocol specifications. A →*net-wide broadcast* may be issued if no route can be found with a limited broadcast.

Route Reply: The answer to a *route request*, destined for the source of the request. The route will be set up during the travel of the route reply or can be carried directly in the *route reply* packet. This is a unicast message. It is abbreviated as RREP, RPY or REPLY in the most common routing protocol specifications.

Route Error: This message-type indicates a broken, stale or otherwise unusable route. It is emitted from the node, which detected the broken route and can be unicast or broadcast. It is abbreviated as RERR or ERROR.

Scenario: See the *Scenario* entry in appendix A.2.

Scope: A term from FSR (cf. Section 8.14). A scope is defined as the set of nodes within a certain distance (in number of hops) from a particular node. Within each scope a different update policy for routing information applies.

Sensor Network: An ad hoc network of tiny sensoric nodes, that are deployed in the target area. The measured data is transmitted in an ad hoc fashion to some collecting node. Sensor networks are rather static, but have only limited transmission ranges and even more limited power capacity.

Sequence Numbers: DSDV (cf. Section 8.9) introduced destination sequence numbers for routes in the routing table. The sequence numbers prevent outdated or stale routes from being entered into the routing table. Many other routing protocols have adapted this method.

Source Tree: A topology graph, representing the current routes from a source to any destination. *Source Trees* are used in several link state protocols, most notably STAR (cf. Section 8.25).

Terminode: A term for the combination of a *terminal* and *node*, which is the common member of ad hoc networks. It was created by the founders of the *Terminode Project* [154], a long term ad hoc WAN project at the EPFL in Switzerland.

Time To Live: See →*Maximum Hop count.*

Virtual Home Region: The *Position Service* suggested for *Terminode routing.*

Virtual Link: In hierarchical routing methods, links on higher layers, than the bottom (physical) layer are called *virtual links*. The need to be mapped to a set of physical links on the bottom layer for the actual communication.

Zone: Some area or set of nodes, that interact in a certain way. In ZRP, a *zone* is defined by the zone radius in terms of hops, in DDR a *zone* covers a tree in the forest routing topology.

Appendix B

GPS Empirical Data

The following data was collected during attempt to gather empirical data with a GPS receiver. I carried a GPS device about two weeks during Summer 2001. I have collected 2500 data samples during that time. This data was not used in any experiment but shows, that it is possible, but very time-consuming to gather movement data to be used in scenarios.

B.1 Location Trace

The figures B.1 and B.2 show a coordinate trace of myself during the sample period. Figure B.2 does not include the far trip. The origin is located near the city center of Munich.

B.2 Direction Changes

Figure B.3 shows the changes in direction from $[-180° : 180°]$ and figure B.4 shows the frequency of direction changes with respect to the angle. Angle intervals have been used, and the direction changes within each interval was accumulated.

B.3 Speed Changes

Figure B.5 shows the speed measured at each sample point in m/s. The speed distribution can be found in figure B.6.

B.4   Distance Distribution

Figure B.7 shows the distance distribution between each samples move. This is influences by the resolution and sample collection algorithm of the GPS device and should be considered with care. The distances are given in m.

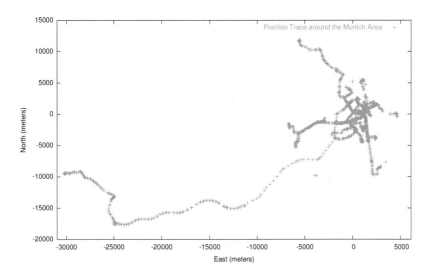

Figure B.1: Location Trace around Munich

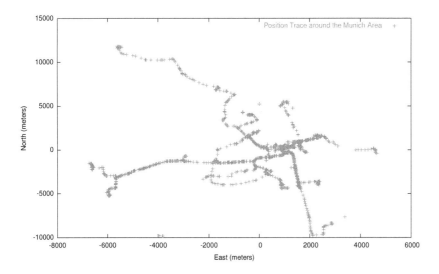

Figure B.2: Location Trace around Munich without far trip

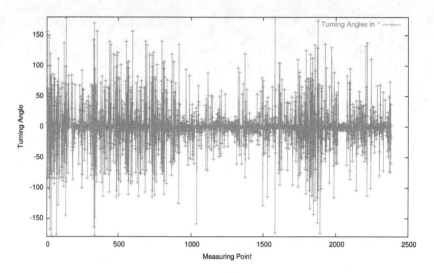

Figure B.3: Direction Changes

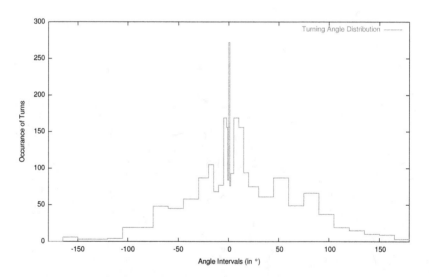

Figure B.4: Distribution of Direction Changes

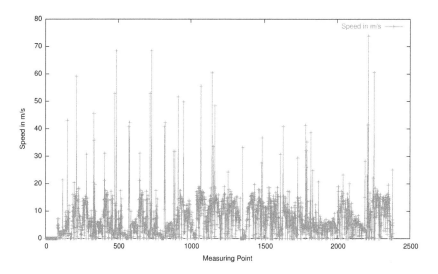

Figure B.5: Speed Changes

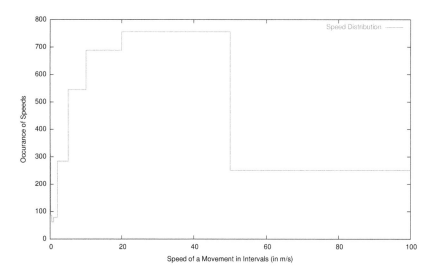

Figure B.6: Speed Distribution

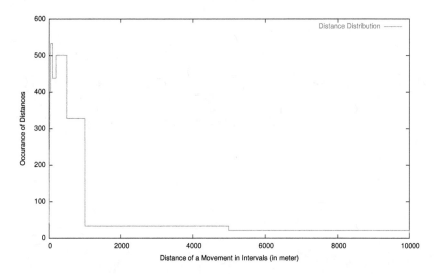

Figure B.7: Distance Distribution

Appendix C

Data from Smooth Model

In order to get some experience with Bettstetter's "Smooth is better than Sharp" Model [15], and also to see if this model can be applied to my GPS collected data (cf. section 6.2 and appendix B), I have implemented the model.

I have used the following parameters:

- $p_{v^*} = 0.04$ (probability of a speed change event)

- $p_{\phi^*} = 0.02$ (probability of a turn event)

	Car	Bicycle	Pedestrian
V_{max} (in m/s)	50.0	7.0	2.0
V_{pref}	0.0 with $P = 0.3$	0.0 with $P = 0.2$	0.0 with $P = 0.2$
	14.0 with $P = 0.3$	4.2 with $P = 0.2$	1.2 with $P = 0.2$
	50.0 with $P = 0.05$	6.5 with $P = 0.1$	2.0 with $P = 0.1$
r_{min} (in m)	8.0	2.0	0.5
a_{max} (in $\frac{m}{s^2}$)	2.5	1.5	2.0
a_{min} (in $\frac{m}{s^2}$)	−4.0	−1.8	−2.0
μ (friction coefficient)	0.7	0.5	0.9

C.1 Distributions of Turns, Distance and Speed

Of each node type (car, bicycle and pedestrian), 30 nodes have been created and modeled. Thus the mean distribution values and standard deviation is based on 30 experiments. For the example results, only a single node of each type is shown.

C.1.1 Direction Change Distribution

The figures C.1, C.2 and C.3 show the distribution of direction changes of cars, bicycles and pedestrians.

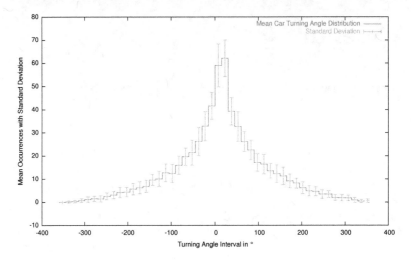

Figure C.1: Distribution of Direction Changes of Cars

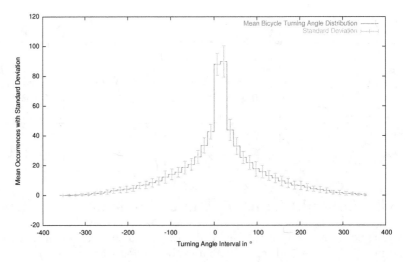

Figure C.2: Distribution of Direction Changes of Bicycles

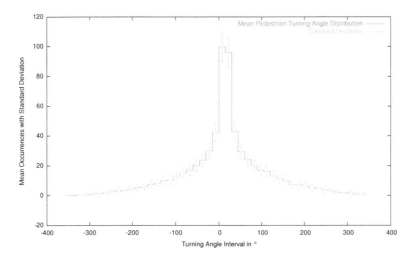

Figure C.3: Distribution of Direction Changes of Pedestrians

C.1.2 Distance Distribution

The figures C.4, C.5 and C.6 show the distribution of distances traveled in each move by cars, bicycles and pedestrians.

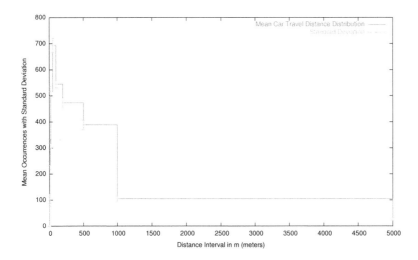

Figure C.4: Distribution of Travel Distances of Cars

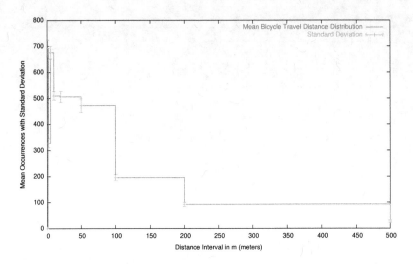

Figure C.5: Distribution of Travel Distances of Bicycles

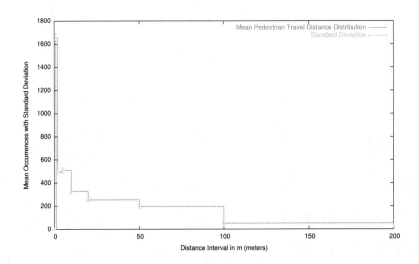

Figure C.6: Distribution of Travel Distances of Pedestrians

C.1.3 Speed Distribution

The figures C.7, C.8 and C.9 show the distribution of speed for each move by cars, bicycles and pedestrians.

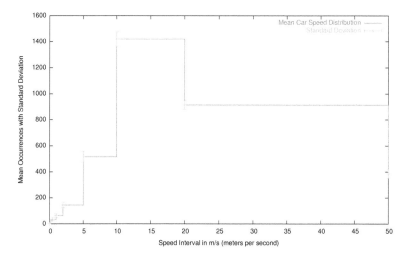

Figure C.7: Distribution of Speed of Cars

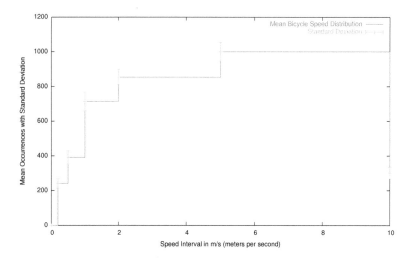

Figure C.8: Distribution of Speed of Bicycles

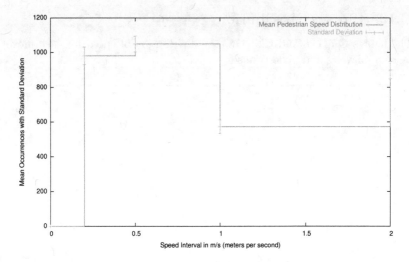

Figure C.9: Distribution of Speed of Pedestrians

C.2   Example node results

This section shows some example graphs of individual nodes including location traces.

C.2.1   Location Traces

The figures C.10, C.11 and C.12 show the locations of an example node of type car, bicycle and pedestrian each.

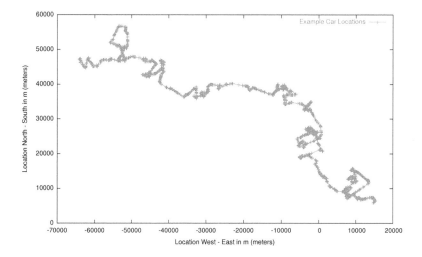

Figure C.10: Locations of a Car

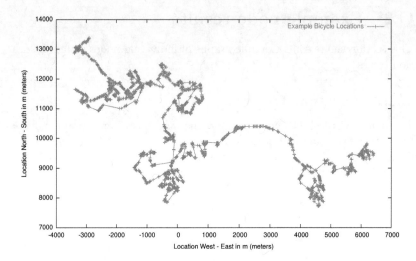

Figure C.11: Locations of a Bicycle

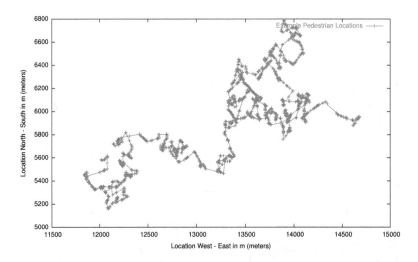

Figure C.12: Locations of a Pedestrian

C.2.2 Directions and Direction Changes

The figures C.13, C.14 and C.15 show the directions of each example node, while the figures C.16, C.17 and C.18 show their turns.

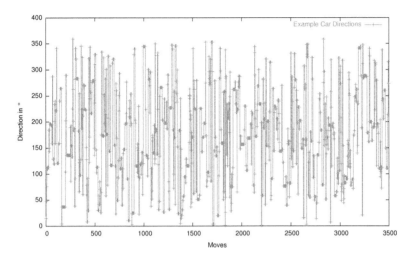

Figure C.13: Directions of a Car

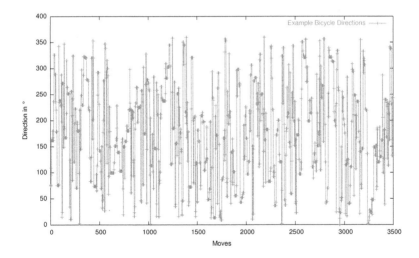

Figure C.14: Directions of a Bicycle

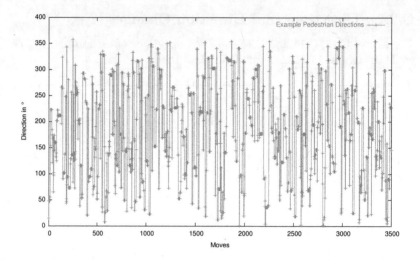

Figure C.15: Directions of a Pedestrian

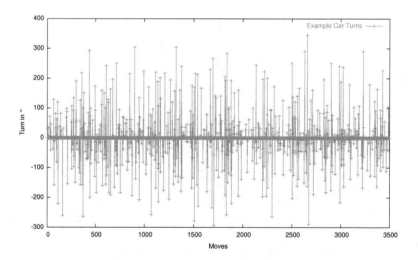

Figure C.16: Direction Changes of a Car

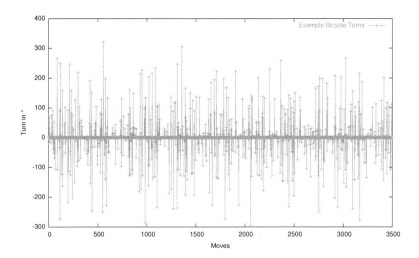

Figure C.17: Direction Changes of a Bicycle

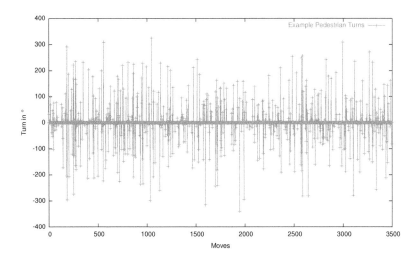

Figure C.18: Direction Changes of a Pedestrian

C.2.3   Distances

The figures C.19, C.20 and C.21 show the distances of each move of the example nodes.

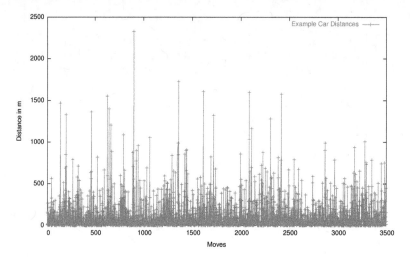

Figure C.19: Distances of Car Movements

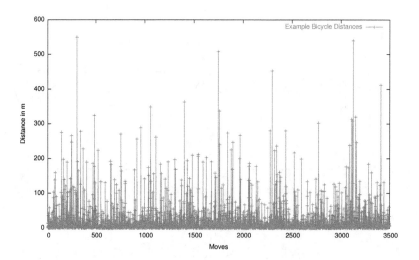

Figure C.20: Distances of Bicycle Movements

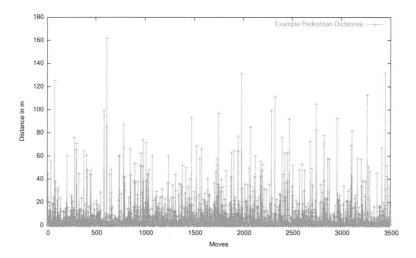

Figure C.21: Distances of Pedestrian Movements

C.2.4 Speed and Speed-changes

The figures C.22, C.23 and C.24 show the speed at the beginning of each movement of the nodes. If the node did accelerate or decelerate in this movement, the resulting speed is the starting speed of the next move. The figures C.25, C.26 and C.27 illustrate this change of speed during a move. This is not the real acceleration (or deceleration), since the movements do not have a fixed duration. The duration of the movements is not shown.

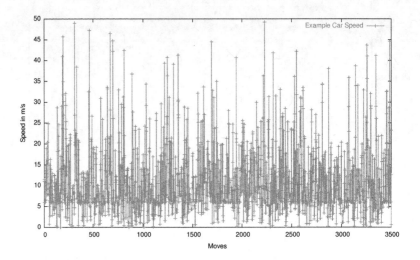

Figure C.22: Speed of Car Movements

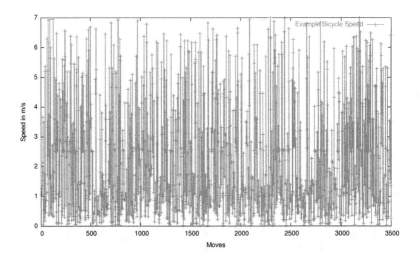

Figure C.23: Speed of Bicycle Movements

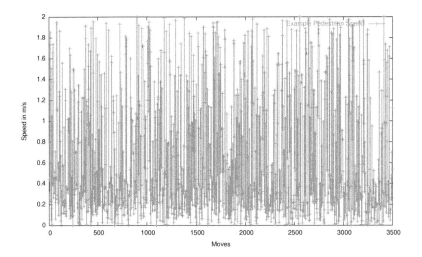

Figure C.24: Speed of Pedestrian Movements

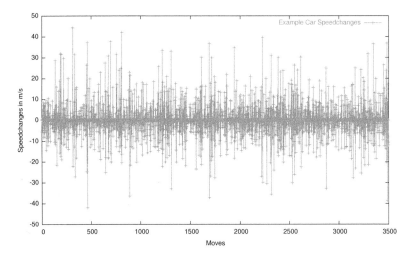

Figure C.25: Change of Speed of Car Movements

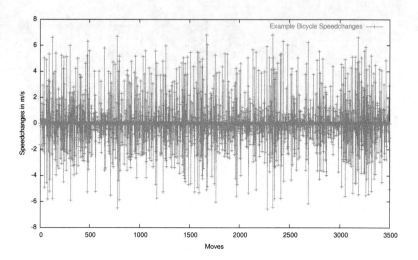

Figure C.26: Change of Speed of Bicycle Movements

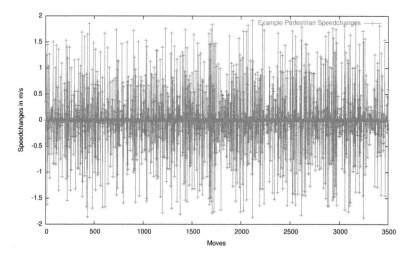

Figure C.27: Change of Speed of Pedestrian Movements

Appendix D

Sample Configuration for Scenario Generator

These two samples files are taken from the scenario generator implemented by Jan Lange[97]. They serve as an example how to specify more realistic scenarios for MANET evaluation.

```
<scenario>
  <name>Katrastrophen Szenario</name>

  <simulationtime>800</simulationtime>
  <width>5000</width>
  <height>2000</height>

  <typedefinition> <!-- globaler Bereich der Typdefinitionen -->

    <nodetypedefinition>  <!-- Knotentypen -->
      <name>Person</name>
      <maxspeed>5.0</maxspeed>
      <minspeed>1.0</minspeed>
      <speedvariation>30.0</speedvariation>
      <maxacceleration>1.0</maxacceleration>
      <minacceleration>5.0</minacceleration>
      <curveRadius>4</curveRadius>
      <nodeattraction name="Fahrzeug">1</nodeattraction>
    </nodetypedefinition>

    <nodetypedefinition>
      <name>Fahrzeug</name>
```

```
        <maxspeed>10.0</maxspeed>
        <minspeed>6.0</minspeed>
        <maxacceleration>3.0</maxacceleration>
        <minacceleration>1.0</minacceleration>
        <speedvariation>60.0</speedvariation>
        <curveRadius>10</curveRadius>
    </nodetypedefinition>

    <nodetypedefinition>
        <name>Hubschrauber</name>
        <maxspeed>50.0</maxspeed>
        <minspeed>30.0</minspeed>
        <maxacceleration>10.0</maxacceleration>
        <minacceleration>10.0</minacceleration>
        <speedvariation>150</speedvariation>
        <curveRadius>30</curveRadius>
        <nodeattraction name="Hubschrauber">-20.0</nodeattraction>
    </nodetypedefinition>


    <!-- Areas/Bereiche -->


    <areatypedefinition>
        <name>Truemmer Typ</name>

        <noderelation name="Person">
            <attraction>800.0</attraction>
            <forbidden>yes</forbidden>
        </noderelation>

        <noderelation name="Fahrzeug">
            <attraction>100.0</attraction>
            <forbidden>yes</forbidden>
        </noderelation>

    </areatypedefinition>

    <areatypedefinition>

        <name>Sammelpunkt Typ</name>
```

```xml
      <noderelation name="Person">
        <attraction>10.0</attraction>
      </noderelation>

      <noderelation name="Fahrzeug">
        <attraction>8.0</attraction>
        <maxspeed>3.0</maxspeed>
      </noderelation>

      <noderelation name="Hubschrauber">
        <attraction>2.0</attraction>
        <maxspeed>5.0</maxspeed>
      </noderelation>

    </areatypedefinition>

  </typedefinition>


<!-- Scenario -->

  <scenariodefinition>

    <area>
      <name>Truemmer 1</name>
      <type>Truemmer Typ</type>
      <rectangle>0, 0, 50, 100</rectangle>
    </area>

    <area>
      <name>Truemmer 2</name>
      <type>Truemmer Typ</type>
      <rectangle>150, 150, 250, 400</rectangle>
    </area>

    <area>
      <name>Truemmer 3</name>
      <type>Truemmer Typ</type>
      <rectangle>1000, 1600, 2000, 2000</rectangle>
    </area>
```

```
<area>
  <name>Truemmer 4</name>
  <type>Truemmer Typ</type>
  <rectangle>800, 1000, 1200, 1300</rectangle>
</area>

<area>
  <name>Truemmer 5</name>
  <type>Truemmer Typ</type>
  <rectangle>2300, 0, 3000, 400</rectangle>
</area>

<area>
  <name>Truemmer 6</name>
  <type>Truemmer Typ</type>
  <rectangle>4800, 0, 5000, 500</rectangle>
</area>

<area>
  <name>Truemmer 7</name>
  <type>Truemmer Typ</type>
  <rectangle>3200, 1100, 4000, 1500</rectangle>
</area>

<area>
  <name>Truemmer 8</name>
  <type>Truemmer Typ</type>
  <rectangle>4800, 1600, 5000, 1700</rectangle>
</area>

<area>
  <name>Truemmer 9</name>
  <type>Truemmer Typ</type>
  <rectangle>900, 400, 1300, 600</rectangle>
</area>

<area>
  <name>Truemmer 10</name>
  <type>Truemmer Typ</type>
  <rectangle>2300, 900, 2700, 1100</rectangle>
```

```
</area>

<area>
  <name>Truemmer 11</name>
  <type>Truemmer Typ</type>
  <rectangle>4000, 200, 4300, 700</rectangle>
</area>




<area>
  <name>Sammelpunkt 1</name>
  <type>Sammelpunkt Typ</type>
  <rectangle>300, 0, 400, 50</rectangle>
  <attractionpoint>310, 30</attractionpoint>
  <attractionpoint>390, 40</attractionpoint>
  <attractionpoint>350, 10</attractionpoint>
</area>

<area>
  <name>Sammelpunkt 2</name>
  <type>Sammelpunkt Typ</type>
  <rectangle>0, 1600, 100, 1800</rectangle>
  <attractionpoint>10, 1750</attractionpoint>
  <attractionpoint>40, 1610</attractionpoint>
  <attractionpoint>30, 1700</attractionpoint>
</area>
<area>
  <name>Sammelpunkt 3</name>
  <type>Sammelpunkt Typ</type>
  <rectangle>4500, 1850, 4900, 2000</rectangle>
  <attractionpoint>4600, 1860</attractionpoint>
  <attractionpoint>4750, 1980</attractionpoint>
  <attractionpoint>4890, 1900</attractionpoint>
</area>



<setnodes>
  <type>Person</type>
```

```
      <number>10</number>
      <rectangle>0,400,800,2000</rectangle>
    </setnodes>
    <setnodes>
      <type>Person</type>
      <number>10</number>
      <rectangle>1300,0,2300,1600</rectangle>
    </setnodes>
    <setnodes>
      <type>Person</type>
      <number>10</number>
      <rectangle>2300,1100,3200,2000</rectangle>
    </setnodes>
    <setnodes>
      <type>Person</type>
      <number>10</number>
      <rectangle>3000,0,4000,1100</rectangle>
    </setnodes>
    <setnodes>
      <type>Person</type>
      <number>10</number>
      <rectangle>4000,500,5000,1600</rectangle>
    </setnodes>

    <setnodes>
      <type>Fahrzeug</type>
      <number>3</number>
      <rectangle>450,0,2000,500</rectangle>
    </setnodes>

    <setnodes>
      <type>Fahrzeug</type>
      <number>4</number>
      <rectangle>50,1600,1000,2000</rectangle>
    </setnodes>

    <setnodes>
      <type>Fahrzeug</type>
      <number>3</number>
      <rectangle>3200,1600,4500,2000</rectangle>
    </setnodes>
```

```
  <setnodes>
    <type>Hubschrauber</type>
    <number>1</number>
    <rectangle>300, 0, 400, 50</rectangle>
  </setnodes>

  <setnodes>
    <type>Hubschrauber</type>
    <number>1</number>
    <rectangle>0, 1600, 100, 1800</rectangle>
  </setnodes>

  <setnodes>
    <type>Hubschrauber</type>
    <number>1</number>
    <rectangle>4500, 1850, 4900, 2000</rectangle>
  </setnodes>

 </scenariodefinition>

</scenario>

#
# Configuration file for a scenario
#
# Comments are made with #, empty lines are ignored.
# Any parameter must take place in its own line and have the form:
# "name" = "value" spaces at any place are ignored
#


# Specifies a number that is used to initialize the generators
# random processor. If not set, a random number will be used.
# The same number here with exactly the same scenario should always
# produce exactly the same movements
# valid is any integer number
GeneratorRandom = 100

# Specifies the number of elements, the field is divided in.
# Can be overridden by the command line option -f
FieldNumber = 200
```

```
# Determines how many fields with the nodes in it in the neighbourhood
# are considered for computing a new movement. That means that for all t
# nodes in them the attraction towards them is figured.
# Minimum is 1 which means that 1 field in each direction is
# considered, giving 9 fields (including the own field of the node).
# A value of 2 would result in 25 fields and so on. Only the existing
# fields are considered so that it does not matter if this number is
# set to high or the "FieldNumber" is to small.
# A value smaller than 1 is converted to 1. The default value is 2.
NodeSurrounding = 3

# Determines how many fields with areas in it in the neighbourhood
# are considered for computing a new movement. Same like
# "NodeSurrounding", so see explantion there.
# Value must 1 or bigger. A smaller value is converted to 1. The
# default is 10
AreaSurrounding = 5

# Specifies the length of the four basic direction vectors that are alwa
# included in the claculation of a new movement. The bigger the number
# is, the more randomized the movement becomes and the less the other
# components are considered. Minimum value is 0.01, smaller values
# will be converted to this. The default is 5
BasicVectorLength = 10

# Gives the random component of the calculation of a new movement.
# Must be a number between 0 (excluding) and 1, where 1 means that the
# weight of a directional component is totally randomized.
# Any value not between 0 and 1 will be converted to 0.5 (the default)
ComponentRandom = 0.9

# If the attraction of the area the currently computed node is in
# shall not be considered, this parameter is to be set to "no".
# If set to "yes" or not specified at all, the attraction (if there is
# one at all) will be computed too.
ComputeCurrentArea = yes

# Determines the default length for a movent of a node. Since the
# parameter "speedvariation" is not mandatory, this value is taken for
# nodes without "speedvaration" set. Any value bigger than 1 is
```

```
# allowed. However, this value should be selected carefully because
# very big values force all nodes to move up to the border of the
# scenario if there is a free way. It is recommended to set the
# speedvariation for each nodetype.
# Any value smaller than 1 will be converted to 1. If this parameter
# is not specified at all, the default value is 100
DefaultSpeedVar = 400


# Specifies the amount of speedvariation, that is figured randomly.
# For determining the length of a new move, the speedvariation is
# broken up in a part that is kept fixed and another part, that is
# figured randomly. This value specifies the ratio of both.
# Valid values are between 0 and 1 both inclusive. A value of 1 means,
# that the figured length will be completly randomized as a value
# between zero and 2*speedvariation. A value of 0 means, that no
# random part is used and all nodes with a free way will move exactly
# as long as the amount of speedvariation.
# Values below 0 are converted to 0, those bigger than 1 to 1. If this
# parameter is not specified, a value of 0.5 is taken.
RandomSpeedVar = 0.9


# Specifies the amount of acceleration, that is figured randomly.
# Same as for RandomSpeedVar but for the acceleration. Legal values
# are between 0 and 0.9.
# Values below 0 are converted to 0, those bigger than 0.9
# to 0.9. If this parameter is not specified, a value of 0.5 is taken.
RandomAcceleration = 0.4


# This parameter gives the maximum time before a new movement of a
# node is recalculated. When the new speed of a node is zero or very
# small, it would never reach its calculated distance. Therefore this
# value specifiec the time after which a new calculation is forced.
# If the value is set very high in combination with very small maximum
# speeds, it is possible to create "standing" nodes. If they keep
# standing in an forbidden area however, they are not able to move out
# anymore.
# Valid numbers are all integers above 10. Values smaller than 10 will
# be converted to 10. The default value is 100.
MaximumStandTime = 1000


# Specifies a log file. If the parameter is not stated, the logfile
```

```
# will be created in the same directory as the input file with the
# extension .log instead of any extension of the input file.
# If a file is specified, the filepath must be specified relativly to
# input file. If NULL is specified, no log file at all will be used.
LogFile = NULL


# This parameter sets the output debug level of the terminal. Valid
# numbers are 0 to 4 where 0 is the lowest level with almost no
# information printed and 4 is the highest level. The default value
# is 0.
TerminalOutLevel = 1


# This parameter sets the output debug level of the logfile. Valid
# numbers are 0 to 4 where 0 is the lowest level with almost no
# information printed and 4 is the highest level. The default value
# is 2.
LogFileOutLevel = 3
```

Appendix E

Sample Simulation Results

The result diagrams of the simulations are presented in this appendix. Each section of this appendix chapter presents the result values of one measured value, as described in section 11.2. Each section lists four pairs of result diagrams. Each of the four pairs presents the results from each of the four scenarios used in this order: Town Center, Country Roads, Disaster Area, Nature Park. The left diagram of all these pairs (i.e. the left column) presents the results from the NoMoGen generated scenarios (the complex mobility model), while the diagrams in the right column present the results from the scengen generated scenarios (the simple Random Waypoint model).

For the NoMoGen results errorbars indicating the standard deviation are included in the diagrams, although the sample was rather small (only three simulations). The scengen experiments consisted only of one sample, thus there is no standard deviation.

The Nature Park scenario was particularly challenging and in many cases the routing protocols were not able to establish any connection.

Some combinations of scenarios and routing protocols could not run any successful simulation due to instabilites (Disaster Area scenario with WRP routing protocol), also some result values measured are not available for all routing protocols (Broken Links, Route Setup Ratio) thus some values are missing.

E.1   Routing Overhead

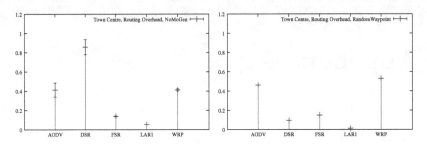

Figure E.1: Routing Overhead - Town Center

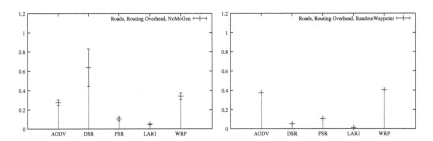

Figure E.2: Routing Overhead - Country Roads

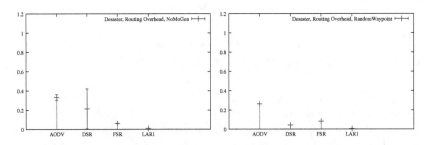

Figure E.3: Routing Overhead - Disaster Area

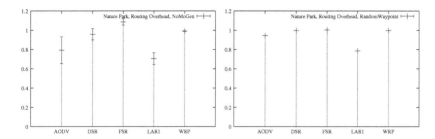

Figure E.4: Routing Overhead - Nature Park

E.2 Route Setup Ratio

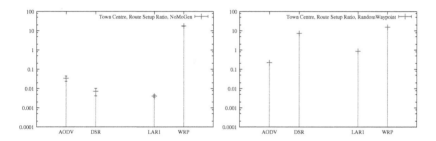

Figure E.5: Route Setup Ratio - Town Center

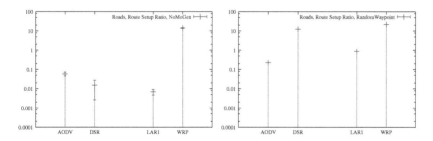

Figure E.6: Route Setup Ratio - Country Roads

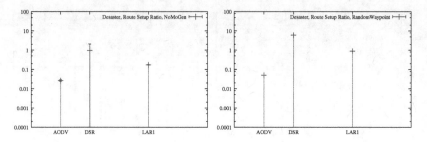

Figure E.7: Route Setup Ratio - Disaster Area

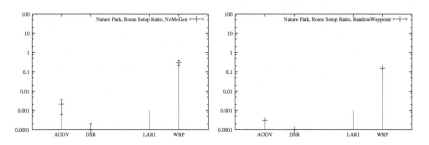

Figure E.8: Route Setup Ratio - Nature Park

E.3   IP Delivery Ratio

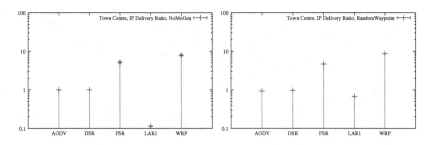

Figure E.9: IP Delivery Ratio - Town Center

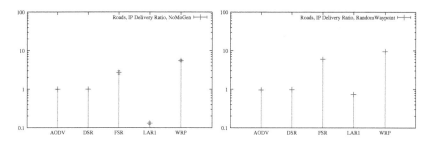

Figure E.10: IP Delivery Ratio - Country Roads

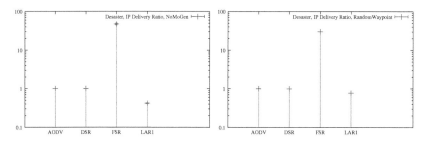

Figure E.11: IP Delivery Ratio - Disaster Area

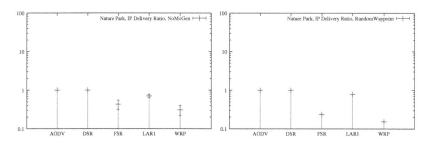

Figure E.12: IP Delivery Ratio - Nature Park

E.4 MAC Broadcast Ratio

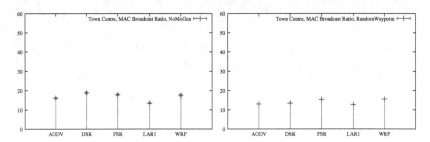

Figure E.13: MAC Broadcast Ratio - Town Center

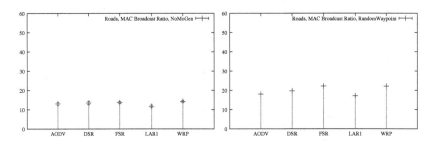

Figure E.14: MAC Broadcast Ratio - Country Roads

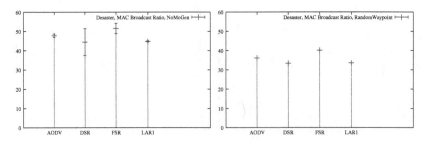

Figure E.15: MAC Broadcast Ratio - Disaster Area

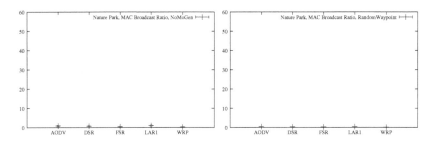

Figure E.16: MAC Broadcast Ratio - Nature Park

E.5 Mean TCP Delay

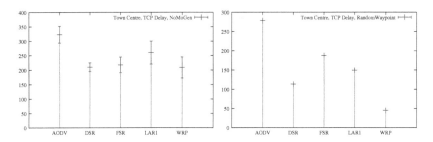

Figure E.17: Mean TCP Delay - Town Center

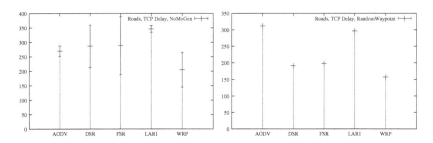

Figure E.18: Mean TCP Delay - Country Roads

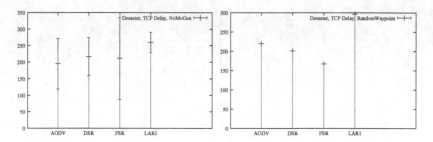

Figure E.19: Mean TCP Delay - Disaster Area

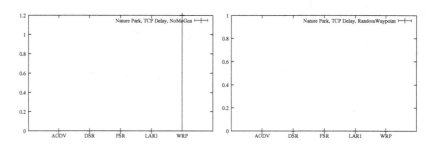

Figure E.20: Mean TCP Delay - Nature Park

E.6   TCP Packets Received

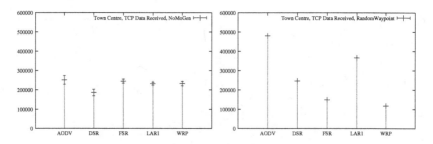

Figure E.21: TCP Packets Received - Town Center

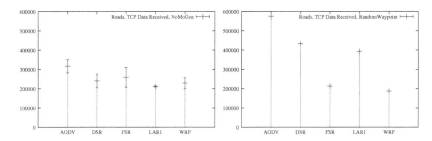

Figure E.22: TCP Packets Received - Country Roads

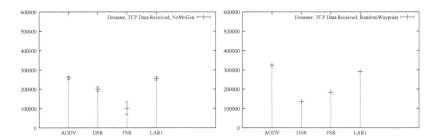

Figure E.23: TCP Packets Received - Disaster Area

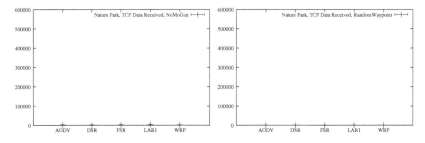

Figure E.24: TCP Packets Received - Nature Park

E.7   TCP Packets Retried

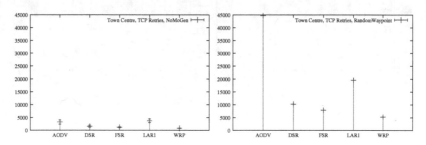

Figure E.25: TCP Packets Retried - Town Center

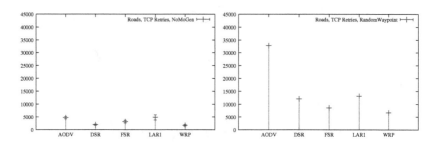

Figure E.26: TCP Packets Retried - Country Roads

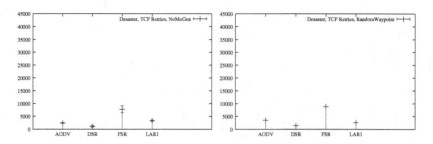

Figure E.27: TCP Packets Retried - Disaster Area

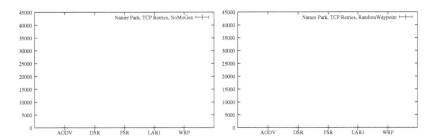

Figure E.28: TCP Packets Retried - Nature Park

E.8 Broken Links

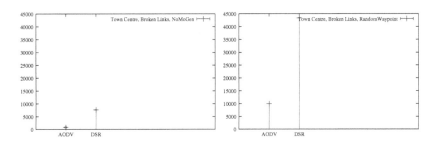

Figure E.29: Broken Links - Town Center

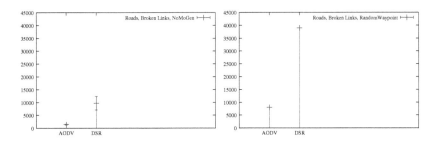

Figure E.30: Broken Links - Country Roads

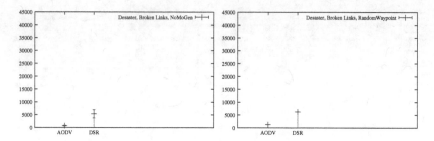

Figure E.31: Broken Links - Disaster Area

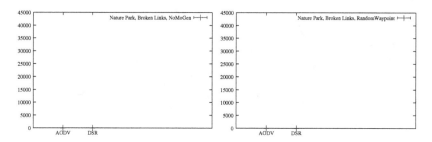

Figure E.32: Broken Links - Nature Park

E.9 FTP Throughput

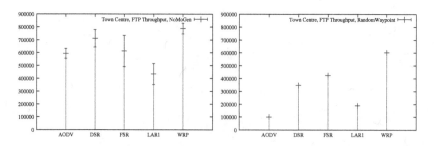

Figure E.33: FTP Throughput - Town Center

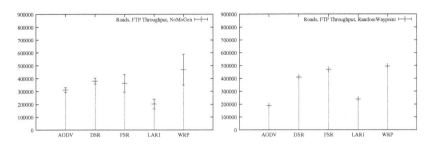

Figure E.34: FTP Throughput - Country Roads

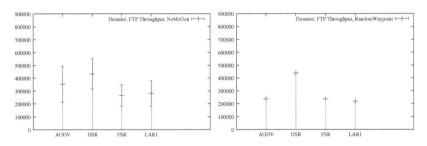

Figure E.35: FTP Throughput - Disaster Area

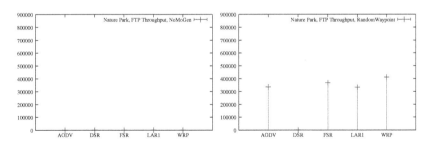

Figure E.36: FTP Throughput - Nature Park

E.10 HTTP Throughput

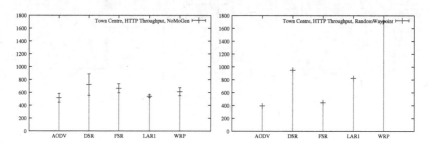

Figure E.37: HTTP Throughput - Town Center

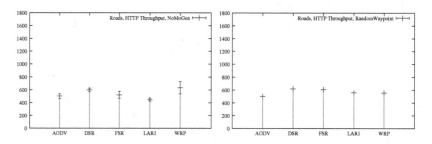

Figure E.38: HTTP Throughput - Country Roads

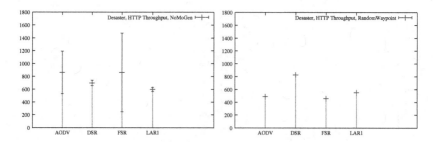

Figure E.39: HTTP Throughput - Disaster Area

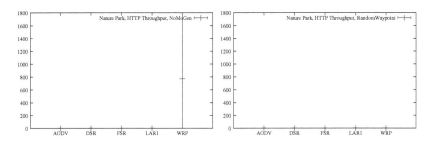

Figure E.40: HTTP Throughput - Nature Park

E.11 TELNET Throughput

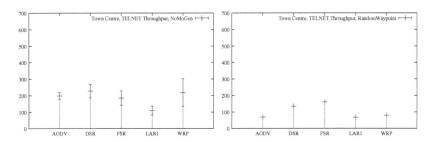

Figure E.41: TELNET Throughput - Town Center

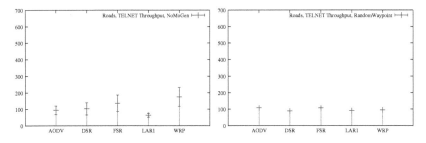

Figure E.42: TELNET Throughput - Country Roads

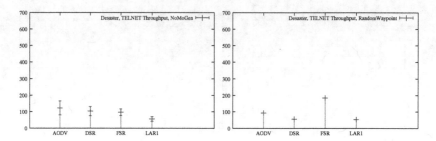

Figure E.43: TELNET Throughput - Disaster Area

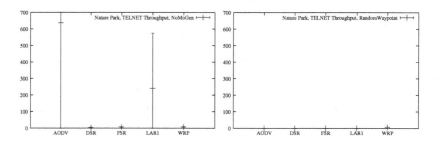

Figure E.44: TELNET Throughput - Nature Park

List of Figures

Bibliography

[1] Elan Amir et al. The network simulator - ns-2. http://www.isi.edu/nsnam/ns/.

[2] George Andreadis. Providing internet access to mobile ad hoc networks, 2002.

[3] Ionut D. Aron and Sandeep K.S. Gupta. A witness-aided routing protocol for mobile ad-hoc networks with unidirectional links. Technical report, Department of Computer Science, Colorado State University, Fort Collins, CO, USA, 1999.

[4] Ionut D. Aron and Sandeep K.S. Gupta. On the scalability of on-demand routing protocols for mobile ad hoc networks: An analytical study. *Journal of Interconnection Networks*, January 2001.

[5] Ionut Dragos Aron. The witness aided routing protocol - a scalable, fault-tolerant solution for mobile ad hoc networks. Master's thesis, Department of Computer Science, Colorado State University, Fort Collins, Colorado, USA, 2000.

[6] S. Bandyopadhyay and E. Coyle. An energy efficient hierarchical clustering algorithm for wireless sensor networks. In *Proceedings of the 22nd Annual Joint Conference of the IEEE Computer and Communications Societies (Infocom 2003)*, 2003.

[7] N. Bansal and Z. Liu. Capacity, delay and mobility in wireless ad-hoc networks. In *Proceedings of the 22nd Annual Joint Conference of the IEEE Computer and Communications Societies (Infocom 2003)*, 2003.

[8] J. Baras and H. Mehta. A probabilistic emergent routing algorithm for mobile ad hoc networks. In *Proc. of WiOpt'03: Modeling and Optimization in Mobile Ad Hoc and Wireless Networks*, March 2003.

[9] Lali Barriere, Pierre Fraigniaud, Lata Narayanan, and Jaroslav Opatrny. Robust position-based routing in wireless ad hoc networks with unstable transmission ranges. In *Proceedings of 5th ACM International Workshop on Discrete Algorithms and Methods for Mobile Computing and Communications (DIALM '01)*, Rome, Italy, July 2001.

[10] Stefano Basagni, Imrich Chlamtac, Violet R. Syrotiuk, and Barry A. Woodward. A distance routing effect algorithm for mobility (dream). In *Proceedings of the ACM/IEEE International Conference on Mobile Computing and Networking, Dallas, TX*, pages 76–84. ACM, October 1998.

[11] Elizabeth M. Belding-Royer. Report on the aodv interop. ucsb tech report 2002-18. Technical report, Dept. of Computer Science, University of California, Santa Barbara, 2002.

[12] Bhargav Bellur and Richard G. Ogier. A reliable, efficient topology broadcast protocol for dynamic networks. In *Proceedings of INFOCOM 1999*, 1999.

[13] M. Benzaid, P. Minet, and K. Agha. Integrating fast mobility in the olsr routing protocol. In *In Proceedings of Fourth IEEE Conference on Mobile and Wireless Communications Networks (MWCN)*, September 2002.

[14] D. Bertsekas and R. Gallager. *Data Networks*, pages 297–333. Prentice-Hall, Englewood Cliffs, N.J., 1987.

[15] Christian Bettstetter. Smooth is better than sharp: A random mobility model for simulation of wireless networks. In *MSWiM'01, ACM International Workshop on Modeling, Analysis and Simulation of Wireless and Mobile Systems*, July 2001.

[16] Christian Bettstetter. Topology properties of ad hoc networks with random waypoint mobility. In *Proceedings of MOBIHOC 2003*, June 2003.

[17] Christian Bettstetter, Hannes Hartenstein, and Xavier Pérez-Costa. Stochastic properties of the random waypoint mobility model. *ACM/Kluwer Wireless Networks*, 2003.

[18] Ljubica Blazevic, Silvia Giordano, and Jean-Yves Le Boudec. A Scalable Routing Method for Irregular Mobile Ad Hoc Networks. Technical report, 2002.

[19] Ljubica Blazevic, Silvia Giordano, and Jean-Yves Le Boudec. A Location Based Routing Method for Mobile Ad Hoc Networks. *IEEE Transactions on Mobile Computing*, 4(2):97–110, 2005.

[20] Ljubica Blazevic, Slivia Giordano, and Jean-Yves Le Boudec. Self-organised wide-area routing. Technical report, EPFL - Communication Systems Department (DSC), Ecole Polytechnique Federale de Lausanne, CH-1015 Lausanne, Switzerland, 2000.

[21] Ljubica Blazevic, Slivia Giordano, and Jean-Yves Le Boudec. Self organized terminode routing. Technical report, EPFL - Communication Systems Department (DSC), Ecole Polytechnique Federale de Lausanne, CH-1015 Lausanne, Switzerland, December 2000.

[22] The bluetooth special interest group. http://www.bluetooth.com/.

[23] Rajendra V. Boppana and Satyadeva Konduru. An adaptive distance vector routing algorithm for mobile ad hoc networks. In *Proceedings of the Twentieth Annual Joint Conference of the IEEE Computer and Communications Societies*, volume 3, pages 1753–1762, 2001.

[24] Linda Briesemeister. *Group Membership and Communication in Highly Mobile Ad Hoc Networks*. Dissertation, Technische Universtiaet Berlin, November 2001.

[25] Josh Broch, David A. Maltz, David B. Johnson, Yih-Chun Hu, and Jorjeta Jetcheva. A performance comparison of multi-hop wireless ad hoc network routing protocols. In *Proceedings of the Fourth Annual ACM/IEEE International Conference on Mobile Computing and Networking, ACM, Dallas, TX*. ACM, October 1998.

[26] Thomas Levin Brunner and Hannes Schmidpeter. Erstellung eines frameworks zur simulation von mobilen ad hoc netzen. Systementwicklungsprojekt, Technische Universität München, July 2003.

[27] S. Buchegger and J. Le Boudec. Performance analysis of the confidant protocol: Cooperation of nodes — fairness in dynamic ad-hoc networks. In *Proceedings of IEEE/ACM Symposium on Mobile Ad Hoc Networking and Computing (MobiHOC)*, June 2002.

[28] Levente Buttyán and Jean-Pierre Hubaux. Nuglets: a virtual currency to stimulate cooperation in self-organized mobile ad hoc networks. Technical report, Institute for Computer Communications and

Applications Department of Communication Systems Swiss Federal Institute of Technology, Lausanne, EPFL-DSC-ICA, CH-1015, Lausanne, Switzerland, January 2001.

[29] Tracy Camp, Jeff Boleng, and Vanessa Davies. A survey of mobility models for ad hoc network research. Technical report, Dept. of Math. and Computer Science, Colorado School of Mines, Golden, CO, April 2002.

[30] Srdan Capkun, Maher Hamdi, and Jean-Pierre Hubaux. Gps-free positioning in mobile ad-hoc networks. In *Proceedings of INFOCOM 2001*, 2001.

[31] Ian D. Chakeres, Elizabeth M. Royer, and Charles E. Perkins. Ietf internet draft: Dynamic manet on-demand routing protocol. http://moment.cs.ucsb.edu/pub/draft-ietf-manet-dymo-03.html, October 2005. Work in progress.

[32] Tsu-Wei Chen and Mario Gerla. Global state routing: A new routing scheme for ad-hoc wireless networks. Technical report, Computer Science Department, University of California, Los Angeles, 1998.

[33] X. Chen, M. Faloutsos, and S. Krishnamurthy. Distance adaptive (dad) broadcasting for ad hoc networks. In *Proceedings of IEEE MILCOM*, 2002.

[34] Yuanzhu Peter Chen, Arthur L. Liestman, and Jiangchuan Liu. *Clustering Algorithms for Ad Hoc Wireless Networks*. Nova Science Publisher, 2004.

[35] Ching-Chuan Chiang, Hsiao-Kuang Wu, Winston Liu, and Mario Gerla. Routing in clustered multihop, mobile wireless networks with fading channel. Technical report, University of California at Los Angeles Computer Science Department, 1997.

[36] Lars Christensen and Gitte Hansen. The optimized link state routing protocol. Master's thesis, Department of Computer Science, Aalborg University, Fredrik Bajers Vej 7, 9220 Aalborg Ost, 2001.

[37] T. Clausen, P. Jacquet, and L. Viennot. Comparative study of routing protocols for mobile ad hoc networks. In *In Proceeding of The First Annual Mediterranean Ad Hoc Networking Workshop*, September 2002.

[38] T. Clausen and Philippe Jacquet. Optimized link state routing protocol (olsr). http://www.ietf.org/rfc/rfc3626.txt, October 2003. Experimental RFC.

[39] M. Scott Corson and Anthony Ephremides. A distributed routing algorithm for mobile wireless networks. *ACM/Balzer Wireless Networks*, 1(1):61–81, February 1995.

[40] M. Scott Corson and Vincent Park. *Link Reversal Routing*, chapter 8, pages 255–298. In Perkins [124], 2001.

[41] Scott Corson. Routing protocol performance issues and evaluation considerations. ftp://ftp.isi.edu/in-notes/rfc2501.txt, 1999.

[42] Y. Dalal and R.Metclafe. Reverse path forwarding of broadcast packets. *Communications of the ACM*, 21(12):1040–1048, December 1978.

[43] Samir R. Das, Robert Castañeda, and Jiangtao Yan. Simulation based performance evaluation of mobile, ad hoc network routing protocols. In *ACM/Baltzer Mobile Networks and Applications (MONET) Journal* [44], pages 179–189.

[44] Samir R. Das, Robert Castañeda, Jiangtao Yan, and Ringli Sengupta. Comparative performance evaluation of routing protocols for mobile, ad hoc networks. In *Proceedings of 7th Int. Conf. on Computer Communications and Networks (IC3N), Lafayette, LA*, pages 153–161. University of Texas at San Antonio, October 1998.

[45] Samir R. Das, Charles E. Perkins, and Elizabeth M. Royer. Performance comparison of two on-demand routing protocols for ad hoc networks. In *Proceedings of INFOCOM 2000 Conference, Tel-Aviv, Israel*, March 2000.

[46] Renato M. de Moraes, Hamid R. Sadjadpour, and J.J. Garcia-Luna-Aceves. On mobility-capacity-delay trade-off in wireless ad hoc networks, 2004.

[47] D.J.Watts. *Small Worlds: The dynamics of networks between order and randomness.* Princeton University Press, 1999.

[48] Rohit Dube, Cynthia D. Rais, Kuang-Yeh Wang, and Satish K. Tripathi. Signal stability based adaptive routing (ssa) for ad-hoc mobile

networks. Technical Report CS-TR-3646, UNMIACS-TR-96-34, Institute for Advanced Computer Studies, Mobile Computing and Multimedia Laboratory, Department of Computer Science, University of Maryland, College Park, MD 20742, August 1996.

[49] Thomas D. Dyer and Rajendra V. Boppana. A comparison of tcp performance over three routing protocols for mobile ad hoc networks. In *Proceedings of the ACM Symposium on Mobile Ad Hoc Networking & Computing (Mobihoc)*. ACM, October 2001.

[50] Thomas D. Dyer and Rajendra V. Boppana. On routing web and multimedia traffic in a mobile ad hoc networks. In *Proceedings of the 36th Hawaii International Conference on System Sciences*, 2003.

[51] Leonardo B. e Oliveira, Isabela G. Siqueira, and Antonio A. F. Loureiro. On the performance of ad-hoc routing protocols under a peer-to-peer application. 65(11):1337–1347, November 2005.

[52] Jakob Eriksson, Michalis Faloutsos, and Srikanth Krishnamurthy. Scalable ad hoc routing: The case for dynamic addressing. In *In Proceedings of IEEE INFOCOM 2004*, 2004.

[53] Laura Marie Feeney. A taxonomy for routing protocols in mobile ad hoc networks. Technical report, Swedish Institute of Computer Science, Box 1263, SE-164 29 Kista, Sweden http://www.sics.se/ lmfeeney/, October 1999.

[54] Holger Füßler, Martin Mauve, Hannes Hartenstein, Michael Käsemann, and Dieter Vollmer. A Comparison of Routing Strategies for Vehicular Ad Hoc Networks. Technical Report TR-02-003, Department of Computer Science, University of Mannheim, 2002.

[55] E. Gafni and D. Bertsekas. Distributed algorithms for generating loop-free routes in networks with frequently changing topology. *IEEE Transactions on Communications*, 29(1):11–15, January 1981.

[56] D. Ganesan, B. Krishnamachari, A. Woo, D. Culler, D. Estrin, and S. Wicker. An empirical study of epidemic algorithms in large scale multihop wireless networks, irb-tr-02-003. Technical report, Intel Research, Berkeley, March 2002.

[57] Jie Gao, Leonidas J. Guibas, John Hershberger, Li Zahng, and An Zhu. A geometric spanner for routing on mobile networks. In Sung Ju Lee, editor, *Mobihoc 2001*, pages 45–55. ACM Sigmobile, 2001.

[58] J.J. Garcia-Luna-Aceves and Marcello Spohn. Scalable link-state internet routing. In *IEEE International Conference on Network Protocols*, October 1998.

[59] J.J. Garcia-Luna-Aceves and Marcelo Spohn. Source-tree routing in wireless networks. In *IEEE International Conference on Network Protocols (ICNP '99)*, October 1999.

[60] J.J. Garcia-Luna-Aceves and Marcelo Spohn. *Bandwidth-Efficient Link-State Routing in Wireless Networks*, chapter 10, pages 323–350. In Perkins [124], 2001.

[61] Mario Gerla, Xiaoyan Hong, and Guangyu Pei. Landmark routing for large ad hoc wireless networks. In *Proceedings of IEEE GLOBECOM 2000*, San Francisco, CA, November 2000.

[62] Mario Gerla, Xiaoyun Hong, and Li Ma. Landmark routing protocol (lanmar) for large scale ad hoc networks. Internet-Draft http://www.ietf.org/internet-drafts/draft-ietf-manet-lanmar-04.txt, June 2002. Work in progress.

[63] Mario Gerla, Guangyu Pei, Xiaoyan Hong, and Tsu-Wei Chen. Fisheye state routing protocol (fsr) for ad hoc networks. Internet Draft http://www.ietf.org/internet-drafts/draft-ietf-manet-fsr-00.txt, November 2000. Work in progress.

[64] Robert S. Gray, David Kotz, Calvin Newport, Nikita Dubrovsky, Aaron Fiske, Jason Liu, Christopher Masone, Susan McGrath, and Yougu Yuan. Outdoor experimental comparison of four ad hoc routing algorithms. In *Proceedings of the ACM/IEEE International Symposium on Modeling, Analysis and Simulation of Wireless and Mobile Systems (MSWiM)*, pages 220–229, October 2004.

[65] Daniel Lihui Gu, Guangyu Pei, Henry Ly, Mario Gerla, and Xiaoyan Hong. Hierarchical routing for multi-layer ad-hoc wireless networks with uavs. In *Proceedings of IEEE Milcom 2000*, 2000.

[66] Manel Guerrero Zapata. Secure ad hoc on-demand distance vector (saodv) routing, September 2005. INTERNET-DRAFT draft-guerrero-manet-saodv-04.txt.

[67] Zygmund J. Haas. A new routing protocol for the reconfigurable wireless networks. Technical report, School of Electrical Engineering, Cornell University, Ithaca, NY, 14853, 1997.

[68] Zygmund J. Haas and Marc R. Pearlman. Determining the optimal configuration for the zone routing protocol. *IEEE Journal on Selected Areas Communications*, 17(8), August 1999.

[69] Zygmund J. Haas and Marc R. Pearlman. The performance of query control schemes for the zone routing protocol. *IEEE/ACM Transactions on Networking*, 9(4):427–438, August 2001.

[70] Zygmunt J. Haas and Marc R. Pearlman. *ZRP A Hybrid Framework for Routing in Ad Hoc Networks*, chapter 7, pages 221–253. In Perkins [124], 2001.

[71] D. Helbing and M. Schreckenberg. Cellular automata simulating experimental properties of traffic flows. *Physical Review*, E59:R2505–R2508, 1999.

[72] Horst Hellbrück. Ansim - adhoc network simulation. http://www.i-u.de/schools/hellbrueck/ansim/.

[73] Gavin Holland and Nitin Vaidya. Analysis of tcp performance over mobile ad hoc networks. In *Proceedings of the Fifth Annual International Conference on Mobile Computing and Networking (MOBICOM)*. Texas A&M University, August 1999.

[74] Yih-Chun Hu, Adrian Perrig, and David B. Johnson. Ariadne: A secure on-demand routing protocol for ad hoc networks. In *Proceedings of the Eighth Annual International Conference on Mobile Computing and Networking* (MobiCom 2002), September 2002.

[75] J. Hubaux, J. Le Boudec, M. Vojnovi'c, S. Giordano, M. Hamdi, L. Blazevi'c, and L. Butty'an. Toward mobile ad-hoc wans: Terminodes. In *Proceedings of the IEEE Wireless Communications and Networking Conference (WCNC'2000), Chicago*, page 5. Ecole Polytechnique Federale de Lausanne, IEEE, September 2000.

[76] J. P. Hubaux, S. Giordano, M. Hamdi, and J. Y. Le Boudec. The terminodes project: Towards mobile ad-hoc wans. In *Proceedings of MOMUC'99, San Diego*. EPFL-ICA, November 1999.

[77] Information science insitute, usc school of engineering. http://www.isi.edu/.

[78] Atsushi Iwata, Ching-Chuan Chiang, Guangyu Pei, Mario Gerla, and Tsu wei Chen. Scalable routing strategies for ad hoc wireless networks.

Technical report, Department of Computer Science University of California, Los Angeles, 1999.

[79] Philippe Jacquet and Anis Laouiti. Analysis of mobile ad-hoc network routing protocols in random graph models. Technical Report 3835, INRIA, Domaine de Voluceau, Rocquencourt, BP 105, 78 153 Le Chesany Cedex, France, 1999.

[80] Philippe Jacquet and Laurent Viennot. Overhead in mobile ad-hoc network protocols. Technical Report 3965, Institut National de Recherche en Informatique et en Automatique, Domaine de Voluceau, Rocquencourt, BP105, 78153, LE CHESNAY Cedex France, June 2000.

[81] Amit Jardosh, Elizabeth M. Belding-Royer, Kevin C. Almeroth, and Subhash Suri. Towards realistic mobility models for mobile ad hoc networks. In *Proceedings of MobiCom*, pages 217–229, San Diego, CA, September 2003.

[82] Amit Jardosh, Elizabeth M. Belding-Royer, Kevin C. Almeroth, and Subhash Suri. Towards realistic mobility models for mobile ad hoc networks. In *Proceedings of MobiCom*, September 2003.

[83] The source for java technology. http://java.sun.com/.

[84] Mingliang Jiang, Jinyang Li, and Y.C. Tay. Cluster based routing protocol(cbrp). Internet Draft ftp://ftp.leo.org/pub/comp/doc/standards/internet-drafts/manet-cbrp-spec/draft-ietf-manet-cbrp-spec-01.txt.gz, August 1999. Work in progress, draft expired.

[85] Per Johansson, Tony Larsson, Nicklas Hedman, Bartosz Mielczarek, and Mikael Degermark. Scenario-based performance analysis of routing protocols for mobile ad-hoc networks. In *Proceedings of The Fifth Annual ACM/IEEE International Conference on Mobile Computing and Networking (MobiCom'99)*, pages 195–206. ACM/IEEE, August 1999.

[86] David B. Johnson and David A. Maltz. Dynamic source routing in ad hoc wireless networks. Technical report, Computer Science Department Carnegie Mellon University, 5000 Forbes Avenue Pittsburgh, PA 15213-3891, 1996.

[87] David B. Johnson, David A. Maltz, and Josh Broch. *DSR The Dynamic Source Routing Protocol for Multihop Wireless Ad Hoc Networks*, chapter 5, pages 139–172. In Perkins [124], 2001.

[88] David B. Johnson, David A. Maltz, Yih-Chun Hu, and Jorjeta G. Jetcheva. The dynamic source routing protocol for mobile ad hoc networks (dsr). http://www.ietf.org/internet-drafts/draft-ietf-manet-dsr-10.txt, 2004. Work in Progress.

[89] Brad Karp and H.T.Kung. Gpsr: Greedy perimeter stateless routing for wireless networks. In *Processings of the 6th Annual ACM/IEEE International Conference on Mobile Computing and Networking (Mobicom 2000)*, pages 243–254, August 2000.

[90] Brad Nelson Karp. *Geographic Routing for Wireless Networks*. Dissertation, Harvard University, Cambridge, Massachusetts, October 2000.

[91] Marc-Olivier Killijian, Raymond Cunningham, Rene Meier, Laurent Mazare, and Vinny Cahill. Towards group communication for mobile participants. Technical report, Trinity College of Dublin, Dublin 2, Eire, 2001.

[92] Felix Ko. Kapazitätssteigerung von multi-hop ad-hoc netzwerken durch power control. Diplomarbeit, Technische Universität München, October 2001.

[93] Young-Bae Ko and Nitin H. Vaidya. Location-aided routing (lar) in mobile ad hoc networks. Technical report, Department of Computer Science Texas A&M University, College Station, TX 77843-3112, June 1998.

[94] Young-Bae Ko and Nitin H. Vaidya. Optimizations for location-aided routing (lar) in mobile ad hoc networks (a brief note). Technical Report 98-023, Department of Computer Science, Texas A&M University, College Station, TX 77843-3112, November 1998.

[95] Jiejun Kong. Anodr: Anonymous on demand routing with untraceable routes for mobile ad-hoc networks. In *Proc. 4th ACM Int. Symposium on Mobile Ad-Hoc Networking and Computing (MobiHoc)*. ACM/SIGMOBILE, June 2003.

[96] Fabian Kuhn, Roger Wattenhofer, and Aaron Zollinger. Worst-Case Optimal and Average-Case Efficient Geometric Ad-Hoc Routing. In

Proc. 4th ACM Int. Symposium on Mobile Ad-Hoc Networking and Computing (MobiHoc), pages 267–278, 2003.

[97] Jan Lange. Implementierung eines szenariogenerators für mobile ad hoc netze. Studienarbeit, Technische Universität München, October 2002.

[98] Tony Larsson and Nicklas Hedman. Routing protocols in wireless ad-hoc networks - a simulation study. Master's thesis, Lulea University of Technology, 1998.

[99] Sung-Ju Lee. *Routing and Multicasting Strategies in Wireless Mobile Ad hoc Networks.* PhD thesis, University of California at Los Angeles, 2000.

[100] Sung Ju Lee, Mario Gerla, and C-K. Toh. A simulation study of table-driven and on-demand routing protocols for mobile ad hoc networks. *IEEE Network*, 13:48–54, 1999.

[101] Jinyang Li, John Jannotti, Douglas S.J. DeCouto, David R. Karger, and Robert Morris. A scalable location service for geographic ad hoc routing. In *Proceedings of the 6th Annual Internation ACM/IEEE Conference on Mobile Computing and Networking (MobiCom 2000)*, August 2000.

[102] Xiang-Yang Li, Gruia Calinescu, and Peng-Jun Wan. Distributed construction of planar spanner and routing for ad hoc wireless networks. Technical report, Illinois Institute of Technology, Chicago, Chigago, IL, 60616, 2001.

[103] G. Lin, G. Noubir, and R. Rajamaran. Mobility models for ad-hoc network simulation. In *In Proceedings of IEEE INFOCOM 2004*, 2004.

[104] Xu Lin and Ivan Stojmenivoc. Gps based distributed routing algorithms for wireless networks. Technical report, Computer Science, SITE, University of Ottawa, Ottawa, Ontario K1N 6N5, Canada, 2000.

[105] Haiyun Luo, Petros Zefros, Jiejun Kong, Songwu Lu, and Lixia Zhang. Self-securing ad hoc wireless networks. In *Seventh IEEE Symposium on Computers and Communications (ISCC '02)*, 2002.

[106] Joseph Macker and Scott Corson. Mobile ad-hoc networks (manet). http://www.ietf.org/proceedings/01dec/183.htm, December 2001. This charter is a snapshot of the 52nd IETF Meeting in Salt Lake City, Utah USA. It may now be out-of-date.

[107] Joseph Macker and Scott Corson. Mobile ad-hoc networks (manet). http://www.ietf.org/html.charters/manet-charter.html, June 2003.

[108] Tatiana K. Madsen, Frank H.P. Fitzek, and Ramjee Prasad. Simulating mobile ad hoc networks: Estimation of connectivity probability, 2004.

[109] Maisie - a c based simulation language. http://may.cs.ucla.edu/projects/maisie/.

[110] David A. Maltz. Resource management in multi-hop ad hoc networks. Technical report, School of Computer Science, Carnegie Mellon University, Pittsburgh, PA 15213, November 1999.

[111] David A. Maltz, Josh Broch, and David B. Johnson. Experiences designing and building a multi-hop wireless ad hoc network testbed. Technical report, The CMU Monarch Project Computer Science Department Carnegie Mellon University, Pittsburgh, PA 15213, 1999. http://www.monarch.cmu.edu/.

[112] S. Manikandan, R. Naveenan, R.K. Padmanaban, and V. Ramachandran. Optimized associativity-based threshold routing for mobile ad-hoc networks. Technical report, College of Engineering, Guindy, Anna University, 2001.

[113] Martin Mauve, Jörg Widmer, and Hannes Hartenstein. A survey on position-based routing in mobile ad-hoc networks. *IEEE Network Magazine*, 15(6):30–39, November 2001.

[114] Shree Murthy and J.J. Garcia-Luna-Aceves. An efficient routing protocol for wireless networks. Technical report, Copmuter Engineering, University of California at Santa Cruz, Santa Cruz, CA 95064, 1996.

[115] Navid Nikaein, Houda Labiod, and Christian Bonnet. Ddr-distributed dynamic routing algorithm for mobile ad hoc networks. Technical report, Institut Eurecom, 2229 Route des Cretes, B.P.193 06904 Sophia Antipolis, France, 2000.

[116] The network simulator ns-2. http://www.isi.edu/nsnam/.

[117] Richard G. Ogier, Fred L. Templin, and M. Lewis. Topology dissemination based on reverse-path forwarding (tbrpf). Experimental RFC ftp://ftp.rfc-editor.org/in-notes/rfc3684.txt, February 2004.

[118] OPNET Technologies, Inc. Opnet modeler. Commercial Product. http://www.opnet.com/products/modeler/home.html.

[119] Manoj Pandey and Daniel Zappala. A scenario-based evaluation of mobile ad hoc multicast routing protocols, July 2004.

[120] P. Papadimitratos, Z. Haas, and E. Sirer. Path-set selection in mobile ad hoc networks. In *Proceedings of 3rd ACM Int. Symposium on Mobile Ad-Hoc Networking and Computing (MobiHoc)*. ACM/SIGMOBILE, June 2002.

[121] Vincent D. Park and M.Scott Corson. A highly adaptive distributed routing algorithm for mobile wireless networks. In *Proceedings of IN-FOCOM 1997*, 1997.

[122] Parsec - parallel simulation environment for complex systems. http://pcl.cs.ucla.edu/projects/parsec/.

[123] Guangyu Pei, Mario Gerla, and Xiaoyan Hong. Lanmar: Landmark routing for large scale wireless ad hoc networks with group mobility. In *roceedings of the ACM/IEEE Workshop on Mobile Ad Hoc Networking and Computing (MOBIHOC), Boston, MA*, pages 11–18. ACM/IEEE, August 2000.

[124] Charles E. Perkins, editor. *Ad Hoc Networking*. Addison-Wesley, 2001.

[125] Charles E. Perkins, Elizabeth M. Belding-Royer, and Samir R. Das. Ad hoc on-demand distance vector (aodv) routing. http://www.ietf.org/rfc/rfc3561.txt, July 2003. Experimental RFC.

[126] Charles E. Perkins and Pravin Bhagwat. Highly dynamic destination-sequenced distance-vector routing (dsdv) for mobile computers. In *ACM SIGCOMM'94 Conference on Communications Architectures, Protocols and Applications*, pages 234–244, 1994.

[127] Charles E. Perkins and Pravin Bhagwat. *DSDV Routing over a Multihop Wireless Network of Mobile Computers*, chapter 3, pages 53–74. In Perkins [124], 2001.

[128] Charles E. Perkins and Elizabeth M. Royer. *The Ad Hoc On-Demand Distance-Vector Protcol*, chapter 6, pages 173–219. In Perkins [124], 2001.

[129] Xavier Pérez-Costa, Christian Bettstetter, and Hannes Hartenstein. Towards a mobility metric for reproducible and comparable results in ad hoc networks research. In *Proceedings of Mobicom 2003*, 2003.

[130] Li Qiming. Scengen. http://www.comp.nus.edu.sg/ liqim-ing/fyp/scengen/. Website of the scenario generator ScenGen, contact: liqiming@comp.nus.edu.sg.

[131] S. Radhakrishnan, Gopal Racherlat, Chandra N. Sekharant, N.S.V. Rao, and Steven G. Batsell. Dst - a routing protocol for ad hoc networks using distributed spanning trees. *IEEE Wireless Communications and Networking Conference*, pages 100–104, 1999.

[132] Jyoti Raju and J.J. Garcia-Luna-Aceves. A comparison of on-demand and table driven routing for ad-hoc wireless networks. In *Proceedings of ICC 2000*, 2000.

[133] Jyoti Raju and J.J. Garcia-Luna-Aceves. Scenario-based comparison of source-tracing and dynamic source routing protocols for ad hoc networks. *ACM Computer Communication Review - Special Issue in mobile extensions to the Internet*, October 2001.

[134] Hartmut Ritter, Thiemo Voigt, Min Tian, and Jochen Schiller. A highly flexible testbed for studies of ad-hoc network behaviour. In *3rd International Workshop on Wireless Local Networks, WLN 2003*, Bonn/Königswinter, Germany, October 2003.

[135] Souma Roy and J.J. Garcia-Luna-Aceves. Using minimal source trees for on-demand routing in ad hoc networks. In *Proceedings if INFO-COM*, pages 1171–1181, 2001.

[136] Elizabeth M. Royer and C.-K. Toh. Review of current routing protocols for ad-hoc mobile wireless networks. Technical report, UCSB, 1999.

[137] Cesar Santivanez. Asymptotic behaviour of mobile ad hoc routing protocols with respect to traffic, mobility and size. Technical report, Center for Communications and Digital Signal Processing, Department of Electrical and Computer Engineering, Northeastern University, Boston, MA, 02115, October 2000.

[138] Cesar Santivanez, Ram Ramanathan, and Ioannis Stavrakakis. Making link-state routing scale for ad hoc networks. In *Proceedings of the 2001 ACM International Symposium on Mobile Ad Hoc Networking and Computing*, pages 22–32. ACM, October 2001.

[139] Cesar A. Santivanez and Ram Ramanathan. Hazy sighted link state (hsls) routing: A scalable link state algorithm. Technical report, Internetwork Research Department, BBN Technologies, 10 Moulton St., Cambridge, MA 02138, August 2001.

[140] Yoav Sasson, David Cavin, and André Schiper. Probabilistic broadcast for flooding in wireless mobile ad hoc networks. In *Proceedings of IEEE Wireless Communications and Networking Conference (WCNC 2003)*, March 2003.

[141] Scalable Network Technologies, Inc. Qualnet. http://www.scalable-networks.com/products/qualnet.php.

[142] Subodh Shah, Edwin Hernandez, and Abdelsalam Helal. Cad-hoc: A cad like tool for generating mobility benchmarks in ad-hoc networks. Technical report, Computer and Information Science and Engineering Department, University of Florida, Gainesville, FL 32611-6120, USA, 2001.

[143] Prasun Sinha, Raghupathy Sivakumar, and Vaduvur Bhafghavan. Enhancing ad hoc routing with dynamic virtual infrastructures. In *Proceedings of the 20. annual joint conference of the IEEE Computer and Communicatin Societies (INFOCOM 2001)*, April 2001.

[144] Raghupathy Sivakumar, Prasun Sinha, and Vaduvur Bharghavan. Cedar: a core-extraction distributed ad hoc routing algorithm. In *IEEE Journal on selected areas in communication*, volume 17. IEEE, August 1999.

[145] Tara Small and Zygmund Haas. The shared wireless infostation model — a new ad hoc networking paradigm (or where there is a whale, there is a way). In *Proc. 4th ACM Int. Symposium on Mobile Ad-Hoc Networking and Computing (MobiHoc)*. ACM/SIGMOBILE, June 2003.

[146] Marcelo Spohn. *Routing in the Internet using Partial Link State Information*. Dissertation, University of California Santa Cruz, September 2001.

[147] Marcelo Spohn and J.J. Garcia-Luna-Aceves. Neighborhood aware source-routing. In *Proceedings of the 2001 ACM International Symposium on Mobile Ad Hoc Networking and Computing*, 2001.

[148] Martha Steenstrup. *Cluster-Based Networks*, chapter 4, pages 75–138. In Perkins [124], 2001.

[149] Mirco Musolesi Stephen. Adaptive routing for intermittently connected mobile ad hoc networks, 2005.

[150] William Su and Mario Gerla. Ipv6 flow handoff in ad hoc wireless networks using mobility prediction. In *Proceedings of the IEEE Global Telecommunications Conference (GLOBECOM), Rio de Janeiro, Brazil*, pages 271–275, Department of Computer Science University of California at Los Angeles, C, December 1999. IEEE, IEEE.

[151] William Su, Sung-Ju Lee, and Mario Gerla. Mobility prediction and routing in ad hoc wireless networks. Technical report, Wireless Adaptive Mobility Laboratory, Computer Science Departmen, University of California, Los Angeles, Los Angeles, CA 90095-1596, 2000.

[152] John Sucec and Ivan Marsic. Clustering overhead for hierarchical routing in mobile ad hoc networks. Technical report, Rutgers University, 2002.

[153] Allen C. Sun. The design and implementation of fisheye routing protocol for mobile ad hoc networks. Master's thesis, Department of Electrical and Computer Science, MIT, May 2002.

[154] Terminodes / mobile information and communication systems. http://www.terminodes.org/.

[155] C-K. Toh and Vasos Vassiliou. *The Effects of Beaconing on the Battery Life of Ad Hoc Mobile Computers*, chapter 9, pages 299–321. In Perkins [124], 2001.

[156] Chai-Keong Toh. Associativity-based routing for ad-hoc mobile networks. Technical report, University of Cambridge Computer Laboratory, Cambridge CB2 3QG, United Kingdom, 1996.

[157] P.F. Tsuchiya. The landmark hierarchy: a new hierarchy for routing in very large networks. *Computer Communication Review*, 18(4):35–42, August 1988.

[158] UCLA Parallel Computing Laboratory and Wireless Adaptive Mobility Laboratory. Glomosim: A scalable simulation environment for wireless and wired network systems. http://pcl.cs.ucla.edu/projects/domains/glomosim.html.

[159] University of California at Los Angeles. Glomosim - global mobile information systems simulation library. http://pcl.cs.ucla.edu/projects/glomosim/, 1999.

[160] University of Maryland. Mars maryland routing simulator. http://www.cs.umd.edu/projects/netcalliper/software.html.

[161] Vint - virtual internetwork testbed. http://www.isi.edu/nsnam/vint/index.html.

[162] Larry Wall. Perl - practical extraction and report language. http://www.perl.org/.

[163] B. Williams and T. Camp. Comparison of broadcasting techniques for mobile ad hoc networks. In *Proceedings of the ACM International Symposium on Mobile Ad Hoc Networking and Computing (MOBIHOC)*, pages 194–205, 2002.

[164] X.Lin and L. Stojmenovic. Geographic distance routing in ad hoc wireless networks. Technical Report TR-98-10, University of Ottawa, SITE, December 1998.

[165] The extensible markup language. http://www.xml.org/.

[166] S.-A. Yang and J.S. Baras. Tora, verification, proofs and model checking, March 2003.

[167] J. Yoon, M. Liu, and B. Noble. Random waypoint considered harmful. In *Proceedings of of INFOCOM. IEEE*, volume 2, pages 1312–1321. IEEE, IEEE Service Center, April 2003.

[168] Yan Yu, Ramesh Govindan, and Deborah Estrin. Geographical and energy aware routing: a recursive data dissemination protocol for wireless sensor networks. Technical Report UCLA/CSD-TR-01-0023, UCLA Computer Science Department, May 2001.

[169] C. Zhu and M. Corson. Qos routing for mobile ad hoc networks. In *Proceedings of IEEE Infocom*, June 2001.

[170] Mattias Östergren. Tcp performance in ad hoc networks. Technical Report T2000:14, Swedish Institute of Computer Science, Box 1263, S-164 29 KISTA, SWEDEN, November 2000.

www.ingramcontent.com/pod-product-compliance
Lightning Source LLC
LaVergne TN
LVHW062304060326
832902LV00013B/2034